自动化测试框架设计

虫师 / 编著

电子工业出版社
Publishing House of Electronics Industry
北京·BEIJING

内 容 简 介

本书分为 13 章。第 1 章介绍了自动化测试框架设计基础。第 2~7 章主要介绍单元测试框架的扩展，包括自动化测试报告设计、数据驱动设计、数据库操作封装设计、随机测试数据设计、命令行工具设计和测试框架扩展功能设计。第 8~11 章主要介绍自动化测试库和自动化测试设计模式，包括 Web UI 自动化测试设计、App UI 自动化测试设计、HTTP 接口自动化测试设计和自动化测试设计模式。第 12 章介绍了自动化测试平台化，以及 Seldom 框架如何为平台化提供支持。第 13 章介绍了自动化测试的 AI 探索。

本书适合自动化测试、软件开发和质量保证领域的开发者，以及各大院校计算机科学和软件工程专业的学生阅读，也适合有一定经验的开发人员参考使用。

未经许可，不得以任何方式复制或抄袭本书之部分或全部内容。
版权所有，侵权必究。

图书在版编目（CIP）数据

自动化测试框架设计 / 虫师编著. -- 北京 : 电子工业出版社, 2024. 11. -- ISBN 978-7-121-49057-6
Ⅰ．TP311.561
中国国家版本馆 CIP 数据核字第 202476J0R9 号

责任编辑：安　娜
印　　刷：山东华立印务有限公司
装　　订：山东华立印务有限公司
出版发行：电子工业出版社
　　　　　北京市海淀区万寿路 173 信箱　邮编：100036
开　　本：787×980　1/16　印张：18.75　字数：420 千字
版　　次：2024 年 11 月第 1 版
印　　次：2024 年 11 月第 1 次印刷
定　　价：89.00 元

凡所购买电子工业出版社图书有缺损问题，请向购买书店调换。若书店售缺，请与本社发行部联系，联系及邮购电话：（010）88254888，88258888。
质量投诉请发邮件至 zlts@phei.com.cn，盗版侵权举报请发邮件至 dbqq@phei.com.cn。
本书咨询联系方式：faq@phei.com.cn。

前言

大约在 2015 年的时候，我负责公司的一个社区论坛项目的 UI 自动化测试工作。在使用 Selenium 编写 UI 自动化测试用例的过程中，我逐渐发现 Selenium 写起来有诸多不便。于是，尝试对 Selenium 的 API 进行二次封装，使其集成 HTMLTestRunner.py 报告，又封装了 unittest 单元测试框架的运行器。最终效果是可以比较方便地编写自动化测试用例。之后，我把项目放到了 GitHub 上，命名为 pyse，并在 GitHub 上断断续续地做着一些更新和维护。整体来看，这是一个比较简单的自动化项目。

2019 年，我对自动化相关技术有了更深刻的理解，也许是因为有更多的空闲时间，我重新关注到 GitHub 上的这个项目，并开始在业余时间对其进行迭代更新。由于"pyse"这个名字在 PyPI 仓库中已被其他项目占用，于是我决定将项目更名为"Seldom"，并成功将其提交到 PyPI 仓库。现在，用户可以非常方便地通过 pip 命令安装和使用该项目。

此后，Seldom 进入了快速迭代阶段。截至本书完稿时，Seldom 的大版本已更新至 3.x。它支持 Web UI 测试、API 测试和 App UI 测试等，已经发展成为一个全功能的自动化测试框架。此外，Seldom 提供了对平台化的支持。Seldom 的社群活跃度很高，社群中几乎每天都有问题讨论。通过社群，我了解到 Seldom 已经在许多公司落地应用。

在开发 Seldom 的过程中，我积累了大量经验。在项目设计层面，我考虑了新功能的引入方式，并控制版本迭代以保持向下兼容；在自动化测试框架设计层面，我考虑了如何封装不同库的 API，以便在满足业务需求的同时，提供更便捷、通用的功能。

我一直未将自己积累的自动化测试框架设计经验编写成书，原因有两方面：一方面，工作比较繁忙；另一方面，空闲时间我都在维护 Seldom 项目。一旦收到社群提出的需求、bug，或者我自己产生了一些新想法，我就会投入时间进行开发，它几乎已经成为我的"第二工作"。

因此，写书的计划一直被搁置，直到 2023 年我才断断续续地开始动笔。成书后，我又花费了大量时间改稿。这导致整个写作周期非常长。当然，这也让我有更多的时间思考书中的内容，只保留了我认为最新、最有价值的部分。

最后，由于水平有限，书中难免有错漏之处。如果发现了错误，希望你能将错误反馈给出版社或我本人，我将感激不尽。再次感谢我的家人，正是因为他们无私的支持和默默的付出，我才能有更多的时间学习和维护开源项目；感谢编辑安娜，没有她的帮助，本书无法顺利出版；感谢拿到这本书的读者，正是你们的支持和鼓励，我才有动力继续编写新书。

虫师

2024 年 10 月

目 录

第 1 章 自动化测试框架设计基础 ... 1
1.1 相关概念对比 ... 1
- 1.1.1 库与框架 ... 1
- 1.1.2 工具与框架 ... 3
- 1.1.3 项目与框架 ... 4

1.2 框架设计基础 ... 5
- 1.2.1 框架是独立的 ... 5
- 1.2.2 框架仅实现通用的功能 ... 5
- 1.2.3 框架应该有清晰的定位 ... 6

1.3 单元测试框架 ... 7
- 1.3.1 测试框架的基本功能 ... 7
- 1.3.2 常见的单元测试框架 ... 7

1.4 unittest 扩展 ... 12
- 1.4.1 实现 unittest 扩展 ... 12
- 1.4.2 使用 unittest 扩展 ... 14

1.5 pytest 扩展 ... 15
- 1.5.1 pytest 扩展方式 ... 16
- 1.5.2 实现 pytest 扩展 ... 18
- 1.5.3 使用 pytest 扩展 ... 20

1.6 构建 Python 包 ... 21

		1.6.1 Poetry	21
		1.6.2 PyPI 仓库	24

第 2 章　自动化测试报告设计 ... 25

- 2.1 实现 jsonrunner ... 25
  - 2.1.1 重写 TestResult 类 ... 26
  - 2.1.2 实现 JSONTestRunner 类 ... 27
  - 2.1.3 编写测试用例 ... 30
- 2.2 HTML 主题模板 ... 33
- 2.3 Jinja 模板引擎 ... 37
  - 2.3.1 Jinja2 的基础使用方法 ... 37
  - 2.3.2 在 Python 中使用 Jinja2 ... 38
- 2.4 基于 unittest 生成 HTML 测试报告 ... 40
  - 2.4.1 修改 runner.py 测试运行器 ... 41
  - 2.4.2 修改 HTML 模板 ... 43
  - 2.4.3 编写测试用例 ... 44

第 3 章　数据驱动设计 ... 46

- 3.1 unittest 的数据驱动扩展 ... 46
  - 3.1.1 parameterized 库 ... 46
  - 3.1.2 DDT 库 ... 48
- 3.2 参数化装饰器 ... 51
  - 3.2.1 实现 data 装饰器 ... 52
  - 3.2.2 实现 dict 数据格式支持 ... 54
- 3.3 参数化数据文件 ... 56
  - 3.3.1 JSON 数据转换 ... 57
  - 3.3.2 YAML 数据转换 ... 59
  - 3.3.3 CSV 数据转换 ... 60
  - 3.3.4 Excel 数据转换 ... 62
  - 3.3.5 实现 file_data 装饰器 ... 64

第 4 章　数据库操作封装设计 ... 66

- 4.1 操作数据库技术 ... 66
  - 4.1.1 数据库驱动 ... 66

4.1.2　ORM 67
　4.2　数据库操作封装 69
　　　4.2.1　封装执行和查询 69
　　　4.2.2　封装增查改删 72

第 5 章　随机测试数据设计 78

　5.1　测试工具介绍 78
　　　5.1.1　Faker 78
　　　5.1.2　Hypothesis 库 80
　5.2　随机测试数据实战 82
　　　5.2.1　随机生成手机号 82
　　　5.2.2　随机生成中文姓名 84
　　　5.2.3　获取在线时间 85

第 6 章　命令行工具设计 88

　6.1　用 Python 实现命令行工具 88
　　　6.1.1　argv 的使用 88
　　　6.1.2　argparse 的使用 89
　　　6.1.3　click 的用法 90
　6.2　命令行工具实战 91
　　　6.2.1　实现自动化测试项目脚手架 92
　　　6.2.2　实现性能工具 94
　　　6.2.3　生成命令行工具 97

第 7 章　测试框架扩展功能设计 100

　7.1　测试用例依赖 100
　　　7.1.1　依赖测试用例结果 100
　　　7.1.2　依赖测试条件 103
　7.2　测试用例分类标签 104
　　　7.2.1　实现分类标签 105
　　　7.2.2　使用分类标签 107
　7.3　使用缓存 109
　　　7.3.1　Redis 的使用 109
　　　7.3.2　LRU Cache 112

	7.3.3 磁盘文件模拟 Cache 113
7.4	实现日志 117
	7.4.1 logging 模块 118
	7.4.2 Loguru 库 121
7.5	自定义异常 123

第8章 Web UI 自动化测试设计 126

8.1	主流 Web 测试库 126
	8.1.1 Selenium 126
	8.1.2 Cypress 127
	8.1.3 Playwright 130
8.2	Selenium API 的二次开发 132
	8.2.1 封装：重命名方法 133
	8.2.2 封装：元素定位和动作整合 133
	8.2.3 封装：独立每个函数 135
	8.2.4 封装：链式调用 136
8.3	Selenium 的断言设计 137
	8.3.1 单元测试框架提供的通用断言 138
	8.3.2 封装 Selenium 断言方法 139
8.4	Selenium 环境管理 143
	8.4.1 Selenium Manager 143
	8.4.2 Docker-Selenium 144

第9章 App UI 自动化测试设计 147

9.1	App 移动自动化测试工具介绍 147
	9.1.1 Android 测试工具 147
	9.1.2 iOS 测试工具 148
	9.1.3 Appium 149
	9.1.4 Airtest Project 150
	9.1.5 Open ATX 150
9.2	Appium 基础 151
	9.2.1 Appium 的安装 151
	9.2.2 Appium 的使用 153
9.3	Appium API 封装 155

	9.3.1 Switch 类	155
	9.3.2 Action 类	158
	9.3.3 FindByText 类	161
	9.3.4 KeyEvent 类	165
9.4	Appium 图像与文字识别	167
	9.4.1 images 插件	168
	9.4.2 Appium OCR 插件	173

第 10 章 HTTP 接口自动化测试设计 … 179

10.1	HTTP 客户端库	179
	10.1.1 requests	179
	10.1.2 HTTPX	180
	10.1.3 aiohttp	180
10.2	HTTP 请求方法集成日志	183
10.3	HTTP 接口测试断言设计	189
	10.3.1 断言基础代码	189
	10.3.2 assertPath()	192
	10.3.3 assertJSON()	196
	10.3.4 assertSchema()	201
10.4	实用功能封装	206
	10.4.1 HTTP 接口检查装饰器	206
	10.4.2 方法依赖装饰器	210
	10.4.3 生成 curl 命令	215
10.5	WebSocket 封装与测试	219
	10.5.1 WebSocket 封装	219
	10.5.2 WebSocket 测试	220

第 11 章 自动化测试设计模式 … 223

11.1	设计模式与开发策略	223
	11.1.1 Page Object 模式	223
	11.1.2 Bot 模式	224
11.2	基于 Page Object 模式的相关库	226
	11.2.1 selenium-page-factory	226
	11.2.2 poium 的基本使用	228

11.2.3　poium 的设计原理 231
11.3　API Object 模式 233
11.3.1　AOM 的设计原理 233
11.3.2　AOM 使用示例 235

第 12 章　自动化测试平台化 239

12.1　自动化测试平台化的基本信息 239
12.1.1　性能测试 239
12.1.2　HTTP 接口自动化测试 240
12.1.3　Web UI 自动化测试 241
12.1.4　App UI 自动化测试 242
12.1.5　自动化测试平台化的优缺点 243
12.2　测试框架与测试平台的整合方案 244
12.2.1　unittest 解析用例 245
12.2.2　对测试用例的收集与运行 246
12.3　SeldomQA 项目 253
12.3.1　Seldom 框架 254
12.3.2　seldom-platform 258

第 13 章　自动化测试的 AI 探索 263

13.1　集成 AI 技术的自动化测试平台 263
13.1.1　基于视觉 AI 技术的自动化检测 263
13.1.2　基于 AI 的自动化测试运行 270
13.2　AIGC 在自动化测试中的应用 278
13.2.1　AI 技术辅助生成自动化测试用例 278
13.2.2　基于 LLM 的代理框架 281

第 1 章 自动化测试框架设计基础

在开始设计自动化测试框架之前,有必要了解框架的基本概念,并了解如何为一个框架命名,以及如何维护、升级和管理版本号。之后,尝试为框架开发一些简单的扩展插件,并发布到 PyPI 仓库。

本章笔者将引领你探索开源项目从开发到发布的全流程。

1.1 相关概念对比

作为一本介绍自动化测试框架设计的书,在介绍什么是"测试框架"之前,我们来看看与"框架"相关的几个概念。

1.1.1 库与框架

相信不少同学对 Martin Fowler 并不感到陌生,他提出的著名的"分层自动化测试""Page Object"等概念,对测试领域产生了深远影响。下面先来看看他对库与框架的解释。

> A library is essentially a set of functions that you can call, these days usually organized into classes. Each call does some work and returns control to the client.

从本质上看,库是一组可以被调用的函数,现在通常被组织成类。每个调用都会执行一些工作,并将控制权返回给客户端。

> A framework embodies some abstract design, with more behavior built in. In order to use it you need to insert your behavior into various places in the framework either by sub classing or by plugging in your own classes. The framework's code then calls your code at these points.

框架体现了一些抽象的设计，并内置了更多的行为。为了使用它，你需要通过子类（或插入自己的类）将你的行为插入框架的各个位置。之后，框架的代码会在这些位置调用你的代码。

另一段解释更加容易理解：

The main difference between a library and a framework is determined by who controls the development process, which is known as inversion of control.

库（Library）与框架（Framework）之间的主要区别取决于由谁来控制开发过程，这也被称为控制反转，如图 1-1 所示。

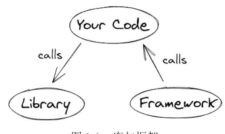

图 1-1　库与框架

如果你深入地使用过一些库和框架，那么应该很认同上面的解释。为了更好地理解库和框架，我们来看一些例子。

（1）requests。

"requests 是一个简单优雅的 Python HTTP 库。"根据官方的定义，毫无疑问，requests 是一个库。

下面这段代码的作用是调用一个 GET 接口。当我们使用 requests 库时，控制权是在我们这里的。当我们想要在程序中发送 HTTP 调用时，可以选择在程序的任意位置使用上面这段代码。

```
import requests
r = requests.get('https://api.git***.com/user', auth=('user', 'pass'))
print(r.status_code)
```

（2）unittest。

"受到 JUnit 的启发，unittest 单元测试框架与其他语言中的主流单元测试框架有着相似的风格。它支持测试自动化，可以共享测试的设置（setUp）和关闭（tearDown）代码。它支持将测试聚合到测试集中，并将测试与报告独立。"

根据官方的介绍我们知道，unittest 是 Python 标准库中集成的单元测试框架。

下面这段代码实现了三条测试用例。

```
import unittest
```

```python
class TestStringMethods(unittest.TestCase):

    def test_upper(self):
        self.assertEqual('foo'.upper(), 'FOO')

    def test_isupper(self):
        self.assertTrue('FOO'.isupper())
        self.assertFalse('Foo'.isupper())

    def test_split(self):
        s = 'hello world'
        self.assertEqual(s.split(), ['hello', 'world'])
        # check that s.split fails when the separator is not a string
        with self.assertRaises(TypeError):
            s.split(2)

if __name__ == '__main__':
    unittest.main()
```

当我们使用 unittest 单元测试框架时，控制权是由框架决定的，根据 unittest 的使用规则：

首先，我们需要创建一个以"test"开头的 Python 文件。

接着，必须创建一个测试类来继承 unittest 的 TestCase 类。

然后，必须以"test"开头定义测试方法（用例），通过 self.assertXXX()使用断言方法。

最后，调用 unittest 的 main()方法运行当前文件中的所有用例。

在整个过程中我们能够感受到 unittest 有许多"规则"需要遵守，否则代码无法正确执行。换言之，框架在控制我们如何写代码。

至此，我们重新看前面的解释，相信你已经有了些许体会。

1.1.2 工具与框架

工具与框架，我们时常会在工作中混用这两个概念，比如 Selenium，有些文章称其为"自动化测试工具"，有些文章称其为"自动化测试框架"，二者有什么不同，下面引用 Quora 网站的一个回答。

> A TOOL is an instrument for special functions in software and system design. It has predefined inputs an delivers predefined outputs.
>
> A FRAMEORK is a wider term. It is an amount of different TOOLs and design procedures in order to create a wider product f. i : a new software system (from analysing of the necessary problem till testing of all evaluated new software) or creating a new software language (including a parser as TOOL).

So a development of a TOOL can be made by an own FRAMEWORK.

A FRAMEWORK includes mental creativity a TOOL needs human factorized usability only.

工具在软件和系统设计中有着特殊的功能，它有预定义的输入，并提供预定义的输出。

框架是一个更广泛的术语。它包含多种工具和设计流程，旨在创建更广泛的产品。比如，创建一个新的软件系统（从分析需求开始到测试所有被评估过的新软件）或创建一种新的软件语言（包括将解析器作为工具）。

因此，我们可以使用框架开发工具。

框架包含了思维的创造性，而工具只需考虑易用性。

根据上面的内容，我们了解到工具更偏向于软件，它可以运行在操作系统之上，并提供独立的交互界面。比如，性能测试工具 LoadRunner。框架是基于编程语言设计出来的。比如，开源自动化测试框架 Robot Framework。

由此抛出一个问题：Selenium 是测试工具还是测试框架？

准确地讲，Selenium 由三部分组成：WebDriver、IDE 和 Grid。其中，WebDriver 是基于不同语言开发出来的用于驱动浏览器的测试库；IDE 是基于 Chrome 或 Firefox 的浏览器插件；Grid 通常被集成到一个 selenium-server.jar 包中。

因此，称 Selenium 为工具或框架都不够准确。Selenium 最核心的部分是 WebDriver，一般我们在使用 Selenium 时，就是在使用 WebDriver 的 API 编写自动化测试用例，如果一定要给 Selenium 分类，那么可以称其为浏览器自动化库。

1.1.3 项目与框架

项目与框架这二者非常容易区分，如果把项目当成一把椅子，那么框架就是制造这把椅子所用到的工具箱。

项目是非常具体的，比如，开发一个用于学习编程的网站或者开发一个用于爬取电影评分的脚本，这些都属于非常具体的项目。

框架是开发这些项目所用到的工具，比如，开发编程网站会用到 Web 框架，开发爬虫脚本会用到爬虫框架。

然而，在软件自动化测试领域，我们更习惯声称自己在"开发自动化测试框架"。

一方面，我们有意无意地被一些技术文章或图书所误导，并未认真地思考二者的区别。

另一方面，我们可以非常容易地将现有的一些框架和库拼装到一起。比如，先将 pytest、Selenium 和 Allure 等集成到一起做 Web UI 自动化测试，然后提供少量的样板用例。准确地讲，这应该被称为"脚手架"或"项目模板"。基于这样的项目模板，我们可以更方便地编写自动化测试用例。

从本质上看，开发自动化测试框架是一种"造轮子"行为，初衷是解决一类问题，并且独立于项目进行开发迭代。接下来，笔者将介绍一个测试框架应该具备哪些特征。

1.2 框架设计基础

想要开发一个框架，就应该先知道框架的三个特点，即框架是独立的、框架仅实现通用的功能，以及框架应该有清晰的定位。

1.2.1 框架是独立的

我们应该把框架当成一个独立的项目来进行开发、维护和升级。

首先，应该为框架起一个独立的名字，既可以以某个动物或植物命名，比如，Python（蟒蛇）或 Lettuce（生菜）；也可以按照框架的本意命名，比如，Robot Framework（自动化框架）或 unittest（单元测试）；还可以是缩写合成词，比如，pytest = Python + Test、Appium= Application + Selenium，关键是简单好记。

其次，框架应该有自己的版本号，推荐使用 GNU 风格的版本号命名。

格式：主版本号.子版本号[.修正版本号[.编译版本号]]。

- 主版本号：重构版本。
- 子版本号：重大功能改进。
- 修正版本号：小升级或者 bug 修复。
- 编译版本号：一般是编译器在编译过程中自动生成的，我们只定义其格式，并不进行人为的控制。

最后，框架应该提供独立的安装。比如，使用 pip 命令进行安装。对于开源的 Python 项目来说，需要创建 setup.py 或 pyproject.toml 文件，即先将项目打包成.whl 格式的文件，再提交到 pypi.org 官方仓库。

1.2.2 框架仅实现通用的功能

作为一款框架，应该解决的是一类问题。以自动化测试框架为例，它应该涵盖定义测试用例、设计测试运行器、确定断言类型、规定测试报告格式，以及实现数据驱动设计等方面。

Robot Framework 是一款非常优秀的自动化测试框架。它独立完成了一款自动化测试框架的所有基础功能，并设计了自己的DSL（Domain-Specific Language，领域特定语言），以及独立的IDE - Robot Framework-ride，建议读者学习和研究它。

相反，框架不应该解决具体的问题。比如，封装某系统的登录功能，有些登录功能是用户名+密码登录、有些登录功能是手机号+验证码，有些登录功能是第三方账号。登录功能应该在自动化测试项目中根据具体功能进行封装，不应该由框架层面实现。

1.2.3 框架应该有清晰的定位

设计框架或库的初衷一定是更好地解决某一类问题。在设计之初，我们应该给框架设计一个清晰的目标和定位。

1. 从无到有地解决一类问题

SUnit 在单元测试框架领域具有开创性意义，下面是 SUnit 在 wikipedia 上的介绍。

xUnit是几个单元测试框架的集合名称，它们的结构和功能都源自Smalltalk的SUnit。SUnit是Kent Beck于1998年设计的，是用一种高度结构化的面向对象风格编写的，这种风格很容易适用于Java和C#等语言。在Smalltalk中引入该框架之后，Kent Beck和Erich Gamma将其移植到Java，并获得了广泛的应用，最终在当前使用的大多数编程语言中都获得了支持。其中许多框架的名称都是SUnit的变体，比如，Java中的JUnit、Python中的unittest。

2. 更加简单地解决一类问题

Flask 是一个使用 Python 编写的轻量级 Web 框架，通过它，我们可以只简单地编写几行代码就搭建一个 Web 服务。后来出现的 Sanic、fastapi 等框架都参考了 Flask 的设计风格。

Flask 代码：

```
from flask import Flask

app = Flask(__name__)
@app.route("/")def hello_world():
    return "<p>Hello, World!</p>"
```

3. 提供更加强大且丰富的功能

Django 是一个开放源代码的由 Python 编写的 Web 框架。它采用了 MTV 的框架模式，最初被用于管理 Lawrence 出版集团旗下的一些以新闻内容为主的网站，即 CMS（内容管理系统）软件。

Django 虽然学习成本较高，但是它的功能提供了 ORM（关系对象映射）、Admin 管理系统、模板系统、Cache 系统、表单处理、会话（session）和国际化等，这些功能几乎都是开箱即用的，可以用来实现一个较为复杂的系统。

当然，一款优秀的框架可能同时具备以上所有特点。

本书对设计一个框架应该具备的基础知识进行了简单介绍。设计框架并非易事，尤其是能够被大众认可并广泛应用的框架，它一定在易用性、功能、代码可靠性、持续维护和使用文档等各个方面都比较优秀，希望我们能对框架设计保持一点儿敬畏之心。

1.3 单元测试框架

1.3.1 测试框架的基本功能

测试框架应该具备以下基本功能。

test case（测试用例）：当在程序层面只有类、方法和函数时，如何定义一个测试用例？不同的测试框架有不同的规则。比如，大部分测试框架都将以"test"开头的方法或函数识别为一条测试用例。

test fixture（测试脚手架）：在运行测试用例前后通常需要完成一些前置或后置的工作。比如，在运行测试用例之前需要构造测试数据，在运行测试用例之后需要清除测试数据等，这些工作都可以在 test fixture 中完成。

test suite（测试套件）：框架在运行测试用例之前，需要查找并添加测试用例到一个集合中，我们一般称这个集合为测试套件。

test runner（测试运行器）：测试运行器主要运行测试套件中的用例，并生成日志或报告。

Assert（断言）：断言主要用于检查测试用例的结果是否准确，进而判定成功或失败。

如果具备了以上功能，那么我们就可以将其视为一个测试框架了。

1.3.2 常见的单元测试框架

在 Python 中有许多优秀的单元测试框架，下面进行简要介绍。

1. unittest

受 JUnit 的启发，unittest 单元测试框架与其他语言中的主流单元测试框架有着相似的风格。它支持测试自动化、配置共享和公共代码测试，支持将测试用例聚合到测试集中，并将测试与报告框架独立。

目前，unittest 已被纳入 Python 标准库，如果你安装了 Python，那么可以直接使用它。

使用示例

```
"""
test_sample_1.py
unittest 例子
"""
import unittest

class MyTest(unittest.TestCase):

    def test_case(self):
        self.assertEqual(2 + 2, 4)

if __name__ == '__main__':
    unittest.main()
```

运行结果

```
> python test_sample_1.py
.----------------------------------------------------------------------
Ran 1 test in 0.000s

OK
```

2. QTAF

QTAF 是由腾讯开源的一款测试工具（框架），其设计风格与 unittest 较为相似。QTA 是腾讯公司部门的缩写，我们也可以称其为 Testbase。

Testbase 是所有 QTA 测试项目的基础，它主要提供测试用例的管理和执行、测试结果和报告、测试项目的管理和配置等功能。

用 pip 命令安装 QTAF。

```
> pip install qtaf
```

使用示例

```
"""
test_sample_2.py
QTAF 例子
"""
from testbase.testcase import TestCase

class HelloTest(TestCase):
    """
    第 1 条用例
    """
    owner = "foo"
    status = TestCase.EnumStatus.Ready
    priority = TestCase.EnumPriority.Normal
```

```
        timeout = 1
    def run_test(self):
        """
        测试步骤
        """
        self.start_step("第1个测试步骤")
        self.log_info("hello")
        self.assert_("检查计算结果", 2 + 2 == 4)

if __name__ == '__main__':
    HelloTest().debug_run()
```

运行结果

```
> python test_sample_2.py
============================================================
测试用例：HelloTest ；所有者：foo ；优先级：Normal； 超时：1 min。
============================================================
----------------------------------------
步骤1：第1个测试步骤
INFO: hello
============================================================
测试用例开始时间：2024-02-27 23:28:09
测试用例结束时间：2024-02-27 23:28:09
测试用例执行时间：00:00:0.01
测试用例步骤结果： 1:通过
测试用例最终结果：通过
============================================================
```

QTAF 的运行结果是用中文显示的，这一点对我们来说比较友好。

3. nose

nose 是 Python 中的一个单元测试框架。nose2 是 nose 的继承者，是一个独特的项目，它并不支持 nose 的所有功能。nose2 的目标是扩展 unittest，使测试变得更容易理解。

用 pip 命令安装 nose2。

```
> pip install nose2
```

我们可以直接使用"nose2"命令运行由 unittest 编写的测试用例。参考前文由 unittest 编写的示例：test_sample_1.py。

运行结果

```
> nose2 -v test_sample_1
test_case (test_sample_1.MyTest.test_case) ... ok
```

```
------------------------------------------------------------------
Ran 1 test in 0.000s

OK
```

这里不需要指定自动化测试文件,"nose2"命令按照规则查找当前目录下面的测试文件,通过后面的"-v"参数可以看到更详细的日志输出。"test_sample_1"即 test_sample_1.py 文件,注意,不要加后缀名。

我们也可以编写 nose2 风格的测试用例。

使用示例

```python
"""
test_sample_3.py
nose2 示例
"""
from nose2.tools import params


@params("Sir Bedevere", "Miss Islington", "Duck")
def test_is_knight(value):
    assert value.startswith('Sir')
```

这个示例用到了 nose2 提供的 params 装饰器参数化测试用例,断言参数化数据是否以"Sir"字符串开头。

运行结果

```
> nose2 -v test_sample_3
test_sample_3.test_is_knight:1
'Sir Bedevere' ... ok
test_sample_3.test_is_knight:2
'Miss Islington' ... FAIL
test_sample_3.test_is_knight:3
'Duck' ... FAIL

======================================================================
FAIL: test_sample_3.test_is_knight:2
'Miss Islington'
----------------------------------------------------------------------
Traceback (most recent call last):
  File "D:\github\test-framework-design\01_chapter\test_framework\test_sample_3.py", line 10, in test_is_knight
    assert value.startswith('Sir')
AssertionError


======================================================================
FAIL: test_sample_3.test_is_knight:3
```

```
'Duck'
----------------------------------------------------------------------
Traceback (most recent call last):
  File
"D:\github\test-framework-design\01_chapter\test_framework\test_sample_3.py",
line 10, in test_is_knight
    assert value.startswith('Sir')
AssertionError

----------------------------------------------------------------------
Ran 3 tests in 0.001s

FAILED (failures=2)
```

4. pytest

pytest 框架使编写小型测试变得容易，pytest 可扩展到支持应用程序和库的复杂功能测试。

在 nose2 的项目中有这样一句话："pytest 是一个优秀的测试框架，我们鼓励用户在新项目中使用它。它拥有强大的维护者团队和更大的用户社区。"由此可见，pytest 的优秀已得到同类型项目的认可。

用 pip 命令安装 pytest。

```
> pip install pytest
```

需要说明的是，pytest 可以运行由 unittest 编写的测试用例。

⌛ 运行结果

```
> pytest -v test_sample_1.py
============================ test session starts =============================
platform win32 -- Python 3.11.4, pytest-8.0.2, pluggy-1.4.0 --
C:\Users\fnngj\.virtualenvs\test-framework-design-BFXq49OO\Scripts\python.exe
cachedir: .pytest_cache
rootdir: D:\github\test-framework-design\01_chapter\test_framework
collected 1 item

test_sample_1.py::MyTest::test_case PASSED
[100%]

============================== 1 passed in 0.01s ==============================
```

pytest 同样按照一定的规则默认查找当前目录中的测试用例并执行。使用"-v"参数可以打印更多当前运行环境的信息。

当然，如果使用 pytest，那么我们应该按照它的风格编写测试用例。

使用示例

```
"""
test_sample_4.py
pytest 例子
"""

def test_answer():
    assert 3 + 1 == 5
```

pytest 的风格与 nose 的风格较为类似，允许我们直接创建测试函数。

运行结果

```
> pytest -q test_sample_4.py
F                                                      [100%]
============================ FAILURES ================================
_____ test_answer _____

    def test_answer():
>       assert 3 + 1 == 5
E       assert (3 + 1) == 5

test_sample_4.py:8: AssertionError
======================== short test summary info =========================
FAILED test_sample_4.py::test_answer - assert (3 + 1) == 5
1 failed in 0.05s
```

这里设置断言的数值不相等，因此可以看到 pytest 为我们提供的错误日志。使用 "-q" 参数表示打印测试结果和与错误相关的信息。

1.4 unittest 扩展

由于 unittest 在编写测试用例时需要创建子类来继承 TestCase 类，所以扩展它的 API 可以通过创建子类实现。比如，Django 框架所提供的测试类就继承自 unittest 的 TestCase 类，unittest 类继承如图 1-2 所示。

首先，在子类中扩展 TestCase 类的功能，然后，在具体项目中使用扩展的 TestCase 子类。

1.4.1 实现 unittest 扩展

下面实现 unittest 扩展，我们将这个扩展命名为 unittest-extend。

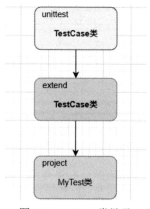

图 1-2　unittest 类继承

📁 目录结构

```
unittest-extend/
├── xtest/
│   ├── __init__.py
│   └── case.py
└── pyproject.toml
```

在 case.py 文件中通过类的继承实现扩展方法。

```python
import time
import random
import unittest


class TestCase(unittest.TestCase):
    """
    unittest 扩展类
    """

    @property
    def get_name(self) -> str:
        """
        随机返回一个英文名
        """
        name_list = ["Andy", "Bill", "Jack", "Robert", "Ada", "Jane", "Eva", "Anne"]
        choice_name = random.choice(name_list)
        return choice_name

    @staticmethod
    def sleep(sec: int = 1) -> None:
        """
        休眠时间
        sec: 秒
        """
        time.sleep(sec)

    def assert_in_text(self, string: str, text: str) -> None:
        """
        将字符串转为小写，断言是否包含字符串
        :param string: 字符串
        :param text: 文本
        """
        self.assertIn(string.lower(), text.lower())


def run(verbosity=1):
    """
    调用 unittest.main() 方法
    """
```

```
unittest.main(verbosity=verbosity)
```

📖 **代码说明**

创建 TestCase 类继承 unittest.TestCase 类。

- get_name()方法用于随机返回一个英文名。
- sleep()方法用于实现休眠。
- assert_in_text()方法用于将字符串转为小写，断言是否包含字符串。

run()函数会将 unittest.main()方法封装之后再给用户使用，这样做的目的是在使用 xtest 时消除 unittest 的导入。

1.4.2 使用 unittest 扩展

安装 unittest-extend 扩展。

```
> git clone https://git***.com/defnngj/unittest-extend    # 克隆项目
> cd unittest-extend                                      # 进入项目目录
> pip install .                                           # 安装
```

创建测试文件 test_sample.py，编写简单的 Selenium 自动化测试代码。

```python
"""
test_sample.py
"""
import xtest
from selenium import webdriver


class MyTest(xtest.TestCase):

    def setUp(self):
        self.driver = webdriver.Chrome()
        self.name = self.get_name

    def tearDown(self):
        self.driver.quit()

    def test_case(self):
        self.driver.get("https://www.ba***.com/")
        search_input = self.driver.find_element("id", "kw")
        search_input.send_keys(self.name)
        search_input.submit()
        self.sleep(2)
        result_title = self.driver.find_elements("css selector", "div > h3.c-title > a")
        for title in result_title:
            print(f"title: {title.text}")
```

```
        self.assert_in_text(self.name, title.text)

if __name__ == '__main__':
    xtest.run()
```

📖 **代码说明**

创建 MyTest 测试类继承 xtest.TestCase 类。整个测试用例是基于 Selenium 实现的一个简单的 Web 自动化测试用例。

- self.get_name 语句可随机生成一个名字，作为搜索关键字。
- self.sleep()方法用于执行休眠方法，不需要 import time 模块。
- self.assert_in_text()方法用于断言，搜索结果的标题是否包含搜索关键字。
- 调用 xtest.run()方法运行测试用例。

⌛ **运行结果**

```
> python test_sample.py

title: andy(阿杜演唱歌曲) - 百度百科
title: andy - Bing 词典
title: Andy - 视频大全 - 高清在线观看
title: andy(韩国神话成员 andy) - 百度百科
title: Andy是什么意思_Andy 的翻译_音标_读音_用法_例句_爱词霸在...
title: Andy是什么意思_Andy 在线翻译_英语_读音_用法_例句_海词词典
title: andy是什么意思? - 百度知道
.
----------------------------------------------------------------------
Ran 1 test in 12.763s

OK
```

扩展 unittest 单元测试框架的功能并不局限于在子类中实现扩展方法，还可以实现测试用例装饰器、测试运行方法，以及命令行工具，我们会在后续的章节详细介绍这些技术。

1.5 pytest 扩展

前面我们介绍了四个 Python 单元测试框架，整体可分为两类：一类是必须有类继承的，比如，QTAF 和 unittest；另一类是可以没有类继承的，比如，nose/nose2 和 pytest。对于可以没有类继承的框架，实现框架的扩展方式会更加多样一些。

1.5.1　pytest 扩展方式

如果想要给 pytest 扩展额外的功能，则可以通过以下几种方式。

1. 钩子函数

利用 conftest.py 文件创建钩子函数。

📂 **目录结构**

```
pytest_extend/
├── conftest.py
└── test_sample.py
```

下面在 conftest.py 文件中定义一个钩子函数 name()，并使用 pytest.fixture 装饰器对其进行装饰。

```python
import pytest

@pytest.fixture
def name():
    return "虫师"
```

pytest.fixture 在单元测试框架中非常重要。这里的 pytest.fixture 默认的级别为 function，可以理解为被装饰的函数会在每个测试用例前被执行。

在 test_sample.py 测试文件中调用钩子函数 name()。

```python
# 调用钩子函数 name()
def test_hello(name):
    greetings = f"hello {name}"
    print(greetings)
    assert greetings == "hello 虫师"
```

在测试用例中，钩子函数 name()作为 test_hello()测试用例的参数被传入。将钩子函数 name()返回的字符串与"hello "进行拼接，并断言拼接的字符串是否为"hello 虫师"。

▷ **运行结果**

```
> pytest -s test_sample.py
========================== test session starts ==========================
platform win32 -- Python 3.11.4, pytest-8.2.2, pluggy-1.5.0
cachedir: .pytest_cache
rootdir: D:\github\test-framework-design\01_chapter\pytest_extend
collected 1 item

test_sample.py hello 虫师
.
=========================== 1 passed in 0.01s ===========================
```

"-s"参数用于设置capture为no，以显示测试用例中print()方法打印的信息。

2. 测试用例装饰器

在pytest中经常会用到参数化装饰器pytest.mark.parametrize来装饰测试用例，用法如下。

```python
import pytest

# 使用装饰器pytest.mark.parametrize
@pytest.mark.parametrize(
    'a, b',
    [
        (1, 2),
        (2, 3),
        (3, 4),
    ])
def test_add(a, b):
    print(f"a:{a}, b:{b}")
    assert a + 1 == b
```

使用pytest.mark.parametrize装饰器，装饰test_add()测试用例，定义a和b为参数变量，每次取一组数据进行测试。

☒ 运行结果

```
> pytest -s test_sample.py
============================ test session starts =============================
platform win32 -- Python 3.11.4, pytest-8.2.2, pluggy-1.5.0
cachedir: .pytest_cache
rootdir: D:\github\test-framework-design\01_chapter\pytest_extend
collected 3 items
test_sample.py a:1, b:2
.a:2, b:3
.a:3, b:4
.
============================ 3 passed in 0.01s ==============================
```

3. 命令行参数

在运行测试用例时可以选择使用"pytest"命令，它提供了许多参数，甚至扩展插件也提供了一些运行参数。下面以pytest的扩展插件pytest-base-url为例进行讲解。

用pip命令安装pytest-base-url。

```
> pip install pytest-base-url
```

编写测试用例如下。

```python
# 使用命令行参数base_url
def test_cmd(base_url):
    print(f"base_url: {base_url}")
    assert "http" in base_url
```

这里用到了命令行参数 base_url，但 base_url 的参数定义是在执行用例时作为参数传入的。

运行结果

```
> pytest -s test_sample.py --base-url https://www.ba***.com
=============================== test session starts ===============================
platform win32 -- Python 3.11.4, pytest-8.2.2, pluggy-1.5.0
baseurl: https://www.ba***.com
rootdir: D:\github\test-framework-design\01_chapter\pytest_extend
plugins: anyio-4.3.0, Faker-24.0.0, base-url-2.1.0
collected 1 item
test_sample.py base_url: https://www.ba***.com
.
=============================== 1 passed in 0.01s ===============================
```

在执行用例后面加上 --base-user 参数，即可指定 URL 地址。

1.5.2　实现 pytest 扩展

我们将这个扩展命名为 pytest-hello。

目录结构

```
pytest-hello/
├── pytest_hello/
│   ├── __init__.py
│   └── plugin.py
└── pyproject.toml
```

核心代码在 pytest_hello.py 文件中，实现的功能如下。

```python
import pytest
from typing import Any, Optional

def pytest_configure(config: Any) -> None:
    """
    register an additional marker
    """
    config.addinivalue_line(
        "markers", "env(name): mark test to run only on named environment"
    )

def pytest_runtest_setup(item: Any) -> None:
    """
    Called to perform the setup phase for a test item.
    """
    env_names = [mark.args[0] for mark in item.iter_markers(name="env")]
    if env_names:
        if item.config.getoption("--env") not in env_names:
            pytest.skip("test requires env in {!r}".format(env_names))
```

```python
@pytest.fixture(scope="function")
def hello(hello_name: str) -> str:
    """
    hello Hook function
    """
    return f"hello, {hello_name}"

@pytest.fixture(scope="function")
def hello_name(pytestconfig: Any) -> Optional[str]:
    """
    hello_name Hook function
    """
    names = pytestconfig.getoption("--hello")
    if len(names) == 0:
        return "虫师"
    if len(names) == 1:
        return names[0]
    return names[0]

def pytest_addoption(parser: Any) -> None:
    """
    Add pytest option
    """
    group = parser.getgroup("hello", "pytest-hello")
    group.addoption(
        "--env",
        action="store",
        default=[],
        help="only run tests matching the environment {name}.",
    )
    group.addoption(
        "--hello",
        action="append",
        default=[],
        help="hello {name}",
    )
```

📖 **代码说明**

在函数 pytest_addoption()中,增加一个命令行组 Hello,并添加两个参数--env 和--hello。

在函数 hello_name()中,通过 pytestconfig 可以获取--hello 参数值。如果为空,则默认值为"虫师";如果为一个值或多个值,则取第 1 个。

在函数 hello()中,获取函数 hello_name()返回的值,并加上"hello,"字符串前缀并返回。

在函数 pytest_configure()中,添加 markers 扩展 env(),用于获取环境名称。

在函数 pytest_runtest_setup()中,获取 markers 中的 env()的值,判断是否等于--env 参数值。

如果不相等，就跳过测试用例，否则执行测试用例。

pytest-hello 项目是采用 pyproject.toml 文件定义安装的，在后面的章节中会介绍如何定义该文件。

1.5.3　使用 pytest 扩展

pytest-hello 的具体安装方式请参考项目的 README.md 文件。

下载并安装 pytest-hello，代码如下。

```
> git clone https://git***.com/defnngj/pytest-hello    # 克隆项目
> cd pytest-hello                                       # 进入项目目录
> pip install .                                         # 安装
```

在安装 pytest-hello 后，通过"pytest --help"命令查看帮助信息。

```
> pytest --help
...

pytest-hello:
    env=ENV           only run tests matching the environment {name}.
    --hello=HELLO     hello {name}...
```

在 test_sample.py 测试文件中，编写测试用例。

```python
import pytest

@pytest.mark.env("test")
def test_case(hello):
    print(f"hello: {hello}")
    assert "hello" in hello
```

pytest.mark.env 装饰器是利用 markers 实现的扩展装饰器。hello()是在 pytest-hello 项目中实现的钩子函数。

☎ 运行结果

不设置--env 参数或设置参数值为非 test，跳过测试用例。

```
> pytest -vs test_sample.py --env dev

collected 1 item
test_sample.py::test_hello SKIPPED (test requires env in ['test'])
```

如果设置了--env 参数值为 test，且未设置--hello 参数，则默认值为"虫师"。

```
> pytest -vs test_sample.py --env test

collected 1 item

test_sample.py::test_case hello: hello, 虫师        PASSED
```

设置--env 参数值为 test，同时设置--hello 参数值为 jack。

```
> pytest -vs test_sample.py --env test --hello jack

collected 1 item

test_sample.py::test_hello hello: hello, jack    PASSED
```

实际上，pytest-hello 这个项目所提供的功能并无多大意义，仅仅用来展示如何开发 pytest 扩展插件。

1.6 构建 Python 包

本章前面介绍了如何实现 unittest 和 pytest 扩展插件，为了让它们能被更多的开发人员使用，我们需要将项目打包并分发。

我们经常使用 pip 命令来安装第三方模块，且整个过程非常简单，这是因为第三方模块的开发人员将他们的项目打包并上传到了 PyPI 官方仓库。

项目打包就是进一步封装项目源代码，并将所有的项目部署工作都事先准备好，这样使用者就可以即装即用，不用操心如何部署的问题。

1.6.1 Poetry

Poetry 是 Python 中用于依赖管理和打包的工具。它允许声明项目所依赖的库，并管理（安装/更新）它们。Poetry 提供了一个锁文件，因此可以重复安装，并且可以构建你的项目用于分发。

Poetry 被看作下一代 Python 包依赖管理和打包工具。下面我们快速学习如何使用 Poetry。

首先，用 pip 命令安装 Poetry。

```
> pip install poetry
```

其次，创建一个新项目，将其命名为 poetry-demo。

```
> poetry new poetry-demo
```

📁 目录结构

```
poetry-demo
├── pyproject.toml
├── README.md
├── poetry_demo
│   └── __init__.py
└── tests
    └── __init__.py
```

其中，最重要的是 pyproject.toml 文件，它的作用与 setup.py 文件类似，主要用于描述项目及其依赖项，内容如下。

```
[tool.poetry]
name = "poetry-demo"
version = "0.1.0"
description = ""
authors = ["fnngj <fnngj@126.com>"]
readme = "README.md"
packages = [{include = "poetry_demo"}]

[tool.poetry.dependencies]
python = "^3.11"

[build-system]
requires = ["poetry-core"]
build-backend = "poetry.core.masonry.api"
```

[tool.poetry]中包含了项目的基本信息，具体如下。

- name：项目名称。
- version：项目版本号。
- description：项目描述，通常只有一句话。
- authors：作者名和邮箱。
- readme：项目描述文件，一般默认为 README.md 文件。
- packages：指定项目的包，其中包含了 poetry_demo 目录，一般在 poetry_demo 目录中实现项目代码。

[tool.poetry.dependencies] 用于定义 Python 版本和第三方库或框架依赖。

- python = "^3.11"：当前项目所依赖的 Python 版本。

[build-system] 用于指定构建系统，这部分不需要修改，默认即可。

- requires：指定 poetry-core，即 Poetry 内核。
- build-backend：构建后端指定的 poetry.core.masonry.api，即 Poetry 的 API。

📁 目录结构

```
pytest-hello/
├── pytest_hello/
│   ├── __init__.py
│   └── plugin.py
├── README.md
└── pyproject.toml
```

以 pytest-hello 项目为例，为它创建 pyproject.toml 文件。

```
[tool.poetry]
```

```toml
name = "pytest-hello"
version = "0.1.0"
description = "pytest hello."
authors = ["fnngj <fnngj@126.com>"]
readme = "README.md"
packages = [{include = "pytest_hello"}]

classifiers = [
    "Programming Language :: Python :: 3",
    "Programming Language :: Python :: 3.8",
    "Programming Language :: Python :: 3.9",
    "Programming Language :: Python :: 3.10",
"Programming Language :: Python :: 3.11",
"Programming Language :: Python :: 3.12",
    "License :: OSI Approved :: Apache Software License",
    "Operating System :: OS Independent",
    "Framework :: Pytest",
]

[tool.poetry.dependencies]
python = ">=3.8"
pytest = "^8.2.2"

[tool.poetry.plugins."pytest11"]
hello = "pytest_hello.plugin"

[build-system]
requires = ["poetry-core"]
build-backend = "poetry.core.masonry.api"
```

安装 pytest-hello 项目。

```
> cd pytest-hello/
> pip install .
Processing d:\github\pytest-hello
  Installing build dependencies ... done
  Getting requirements to build wheel ... done
  Preparing metadata (pyproject.toml) ... done
  ......
```

查看 pytest-hello 项目安装信息。

```
> pip show pytest-hello
Name: pytest-hello
Version: 0.1.0
Summary: pytest hello.
Home-page:
Author: fnngj
Author-email: fnngj@126.com
License:
Location: C:\Python311\Lib\site-packages
Requires: pytest
Required-by:
```

1.6.2 PyPI 仓库

Python Package Index（PyPI）是 Python 编程语言的软件存储库。如果你想让你的项目被更多的 Python 开发人员方便地安装和使用，那么可以选择将项目上传到 PyPI 仓库。

登录 PyPI 网站，注册账号。

Twine 是一个用于在 PyPI 网站上发布 Python 包的程序。它为新项目和现有项目提供了独立于构建系统的源和二进制分发工具。

首先，用 pip 命令安装 Twine。

```
> pip install twine
```

其次，为项目打包，下面基于 pyproject.toml 文件使用 Poetry 工具进行打包。

```
> poetry build
Building pytest-hello (0.1.0)
  - Building sdist
  - Built pytest_hello-0.1.0.tar.gz
  - Building wheel
  - Built pytest_hello-0.1.0-py3-none-any.whl
```

打包完成，在 sdist/目录中会生成.tar.gz 和.whl 两种格式的安装包。

最后，通过 Twine 上传 dist/目录中打包好的项目。

```
> twine check dist/*
> twine upload dist/*
```

📖 **代码说明**

- "check"参数用于检查发行版的长描述是否会在 PyPI 仓库上正确呈现。
- "upload"参数用于将项目上传到 PyPI 仓库。

如果项目仅用于测试，则也可以上传到 TestPyPI 网站。TestPyPI 网站可被认为是测试发布环境，或者叫预发布环境（TestPyPI）。

```
> twine check dist/*
> twine upload -r testpypi dist/*
```

如果想要安装 TestPyPI，则需要指定环境地址，比如：

```
> pip install -i https://test.p***.org/simple/project-name
```

其中，project-name 为待安装的具体项目名。

第 2 章 自动化测试报告设计

这是一个看脸的时代,一个自动化测试工具或框架是否能够吸引用户使用,测试报告的样式和提供的功能非常重要。如果想要为测试框架设计测试报告,应该怎样操作呢?你需要考虑以下两个问题。

(1)如何从测试框架中提取测试数据?比如,每个测试用例的名称、描述和执行结果(成功、失败、错误、跳过)。

(2)如何将测试结果写入 HTML 文件?你需要掌握 HTML、CSS 和 JavaScript 前端开发技术,最好懂一点儿设计,这样才能设计出美观的 HTML 测试报告。

带着上面两个问题,我们正式步入本章的学习之旅。

2.1 实现 jsonrunner

unittest 无法生成 HTML 测试报告,如果想要生成 HTML 测试报告,就需要对它的两个类进行重写,具体如下。

- unittest.TestResult:主要用于记录每个测试用例的结果。
- unittest.TextTestRunner:主要用于运行测试。

HTMLTestRunner 为 unittest 生成 HTML 测试报告提供了模板。后续,我们会看到各种各样定制版本的 HTMLTestRunner。

下面以 HTMLTestRunner 项目为蓝本,学习如何为 unittest 生成 JSON 格式的测试报告。

在 HTMLTestRunner 项目中,HTML 代码和 Python 代码是混合在一起的,而且整个项目的代码都写在一个文件中,阅读起来有一定的难度。为了便于理解,笔者对其进行了简化,只保留了生成 JSON 格式报告的基本功能。

> 📁 **目录结构**

```
02_chapter
├─jsonrunner
│  ├─result.py
│  └─runner.py
├─test_jsonrunner.py
└─result.json
```

2.1.1 重写 TestResult 类

创建 result.py 文件，在文件中创建 _TestResult 类，使其继承 unittest 的 TestResult 类，并重写部分方法。

```python
import sys
from unittest import TestResult

class _TestResult(TestResult):
    """
    继承 unittest 的 TestResult 类，重写成功、失败、错误、跳过等方法
    """
    def __init__(self, verbosity=1):
        TestResult.__init__(self)
        self.success_count = 0
        self.failure_count = 0
        self.error_count = 0
        self.skip_count = 0
        self.verbosity = verbosity
        self.result = []

    def startTest(self, test):
        TestResult.startTest(self, test)

    def addSuccess(self, test):
        self.success_count += 1
        TestResult.addSuccess(self, test)
        self.result.append((0, test, ''))
        if self.verbosity > 1:
            sys.stderr.write('ok')
            sys.stderr.write(str(test))
            sys.stderr.write('\n')
        else:
            sys.stderr.write('.'+str(self.success_count))

    def addFailure(self, test, err):
        self.failure_count += 1
```

```
            TestResult.addFailure(self, test, err)
            _, _exc_str = self.failures[-1]
            self.result.append((1, test, _exc_str))
            if self.verbosity > 1:
                sys.stderr.write('F')
                sys.stderr.write(str(test))
                sys.stderr.write('\n')
            else:
                sys.stderr.write('F')

        def addError(self, test, err):
            self.error_count += 1
            TestResult.addError(self, test, err)
            _, _exc_str = self.errors[-1]
            self.result.append((2, test, _exc_str))
            if self.verbosity > 1:
                sys.stderr.write('E')
                sys.stderr.write(str(test))
                sys.stderr.write('\n')
            else:
                sys.stderr.write('E')

        def addSkip(self, test, reason):
            self.skip_count += 1
            TestResult.addSkip(self, test, reason)
            self.result.append((3, test, reason))
            if self.verbosity > 1:
                sys.stderr.write('S')
                sys.stderr.write(str(test))
                sys.stderr.write('\n')
            else:
                sys.stderr.write('S')
```

📖 **代码说明**

创建 _TestResult 类，使其继承 unittest 的 TestResult 类。

重写 addSuccess()、addFailure()、addError()和 addSkip() 这四个方法，分别用于记录成功、失败、错误和跳过的测试用例，并把每种测试结果都记录到 self.result 列表中。

2.1.2 实现 JSONTestRunner 类

我们需要重写 unittest 的 TextTestRunner 类。从命名可以看出，TextTestRunner 在运行过程中只是简单地将测试结果以文本形式进行打印。现在我们希望把测试结果写入一个 JSON 文件，因此创建了一个名为"JSONTestRunner"的类。

```
import json
import datetime
from jsonrunner.result import _TestResult
```

```python
# 定义用例类型
case_type = {
    0: "passed",
    1: "failure",
    2: "errors",
    3: "skipped"
}

class JSONTestRunner:
    """
    运行测试：生成JSON格式的测试结果
    """

    def __init__(self, output, verbosity=1):
        self.output = output
        self.verbosity = verbosity
        self.start_time = datetime.datetime.now()

    def run(self, test):
        """
        运行测试
        """
        result = _TestResult(self.verbosity)
        test(result)
        case_info = self.test_result(result)

# 将测试结果转为测试报告
        self.result_to_report(case_info)

        stop_time = datetime.datetime.now()
        print(f"Time Elapsed: {stop_time - self.start_time}")
        return result

    def test_result(self, result):
        """
        解析测试结果
        """
        class_list = []
        sorted_result = self.sort_result(result.result)
        for cid, (cls, cls_results) in enumerate(sorted_result):
            # 统计类下面的测试用例数据
            passed = failure = errors = skipped = 0
            for n, t, e in cls_results:
                if n == 0:
                    passed += 1
                elif n == 1:
                    failure += 1
                elif n == 2:
```

```python
                errors += 1
            else:
                skipped += 1

        # 格式化类的描述信息
        if cls.__module__ == "__main__":
            name = cls.__name__
        else:
            name = "%s.%s" % (cls.__module__, cls.__name__)
        doc = cls.__doc__ and cls.__doc__.split("\n")[0] or ""
        desc = doc and '%s' % doc or name

        cases = []
        for tid, (n, t, e) in enumerate(cls_results):
            case_info = self.generate_case_data(cid, tid, n, t, e)
            cases.append(case_info)

        class_list.append({
            "desc": desc,
            "count": passed + failure + errors + skipped,
            "pass": passed,
            "fail": failure,
            "error": errors,
            "skipped": skipped,
            "cases": cases
        })

    return class_list

@staticmethod
def sort_result(result_list):
    """
    unittest 在运行测试用例时没有特定的顺序，
    这里将测试用例按照测试类进行了分组
    """
    rmap = {}
    classes = []
    for n, t, e in result_list:
        cls = t.__class__
        if not cls in rmap:
            rmap[cls] = []
            classes.append(cls)
        rmap[cls].append((n, t, e))
    r = [(cls, rmap[cls]) for cls in classes]
    return r

@staticmethod
def generate_case_data(cid, tid, n, t, e):
    """
    生成测试用例数据
```

```python
    """
    tid = (n == 0 and "p" or "f") + f"t{cid + 1}.{tid + 1}"
    name = t.id().split('.')[-1]
    doc = t.shortDescription() or ""

    case = {
        "number": tid,
        "name": name,
        "doc": doc,
        "result": case_type.get(n),
        "error": e
    }

    return case

def result_to_report(self, result):
    """
    将测试结果转为测试报告
    """
    # 保存为 JSON 文件
    with open(self.output, "w", encoding="utf-8") as json_file:
        json.dump(result, json_file, ensure_ascii=False, indent=2)
```

> 📖 **代码说明**

解析测试用例的过程比较复杂，虽然笔者已经尽力进行了简化，但仍然需要 100 多行代码。

创建 JSONTestRunner 类，在 __init__() 初始化方法中，通过"output"参数接收一个文件，用于保存 JSON 格式的测试结果。

在 run() 方法中，通过"test"参数接收组装测试用例的测试套件，并调用 _TestResult 类记录测试的运行结果。

把结果传给 self.test_result()方法进行解析。整个解析过程主要围绕两个维度：测试类和测试用例。它们分别包含不同的信息。

- 测试类：类名和类描述，统计类下面的测试用例数量（成功、失败、错误和跳过）。
- 测试用例：方法名、方法描述和结果，以及错误信息。

最终，将这些数据组装成列表和字典，调用 result_to_report()方法将其保存到 JSON 文件中。

2.1.3 编写测试用例

通过编写测试用例验证 jsonrunner 的功能是否正确。

```python
import unittest
from jsonrunner.runner import JSONTestRunner
```

```python
class TestDemo(unittest.TestCase):
    """测试 TestDemo 类"""

    def test_pass(self):
        """成功"""
        self.assertEqual(5, 5)

    def test_fail(self):
        """失败"""
        self.assertEqual(5, 6)

    def test_error(self):
        """错误"""
        self.assertEqual(a, 6)

    @unittest.skip("skip case")
    def test_skip(self):
        """跳过"""
        ...


if __name__ == '__main__':
    suit = unittest.TestSuite()
    suit.addTests([
        TestDemo("test_pass"),
        TestDemo("test_skip"),
        TestDemo("test_fail"),
        TestDemo("test_error")
    ])

    runner = JSONTestRunner(output="./result.json")
    runner.run(suit)
```

 代码说明

- 创建 TestDemo 类，实现对测试结果的处理，包括成功、失败、错误和跳过这四种测试用例。
- 通过 TestSuite 类的 addTests() 方法将测试用例添加到测试套件中。
- 调用 JSONTestRunner 类，通过"output"参数指定测试报告存放路径，通过 run() 方法运行测试套件中的测试用例。

 运行结果

打开 result.json 文件，查看运行结果。

```
[
  {
```

```json
    "desc": "Test Demo class",
    "count": 4,
    "pass": 1,
    "fail": 1,
    "error": 1,
    "skipped": 1,
    "cases": [
      {
        "number": "pt1.1",
        "name": "test_pass",
        "doc": "pass case",
        "result": "passed",
        "error": ""
      },
      {
        "number": "ft1.2",
        "name": "test_skip",
        "doc": "skip case",
        "result": "skipped",
        "error": "skip case"
      },
      {
        "number": "ft1.3",
        "name": "test_fail",
        "doc": "fail case",
        "result": "failure",
        "error": "Traceback (most recent call last):\n  File \"D:\\github\\test-framework-design\\02_chapter\\test_jsonrunner.py\", line 14, in test_fail\n    self.assertEqual(5, 6)\nAssertionError: 5 != 6\n"
      },
      {
        "number": "ft1.4",
        "name": "test_error",
        "doc": "error case",
        "result": "errors",
        "error": "Traceback (most recent call last):\n  File \"D:\\github\\test-framework-design\\02_chapter\\test_jsonrunner.py\", line 18, in test_error\n    self.assertEqual(a, 6)\n\nNameError: name 'a' is not defined\n"
      }
    ]
  }
]
```

通过分析 JSON 格式的结果可以看出，在一个[]列表中，通过{}字典记录了每个测试类的结果。在每个测试类中，通过 cases 对应的{}字典记录了每个测试用例的结果。

2.2　HTML 主题模板

从零开始学习 HTML、CSS 和 JavaScript 需要花费不少时间，而且除了这些技术本身，如何设计一套风格统一的测试报告模板也很考验个人的 UI 设计能力，因此最快捷的方式就是使用 HTML 主题模板。Bootstrap 模板如图 2-1 所示。

图 2-1　Bootstrap 模板

Bootstrap 网站提供了许多漂亮的网站主题模板，而想要拿来使用则需要一些技巧，接下来笔者介绍如何快速获取一套网站主题模板。

第 1 步，选择一套你喜欢的主题，单击即可查看详情。Bootstrap 模板地址如图 2-2 所示。打开"前端开发者工具"，找到<iframe>标签的 src 属性，显示真正引用的网址。

第 2 步，复制网址，在新的浏览器窗口中打开该网址，在页面任意位置处单击鼠标右键，在弹出的快捷菜单中单击"查看页面源代码"选项，如图 2-3 所示。

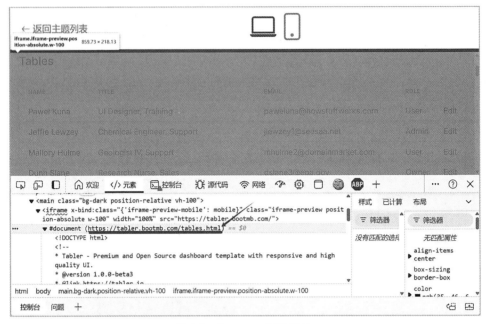

图 2-2 Bootstrap 模板地址

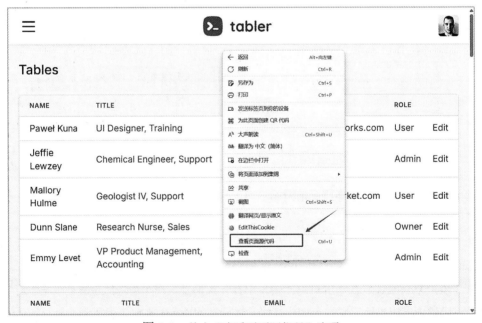

图 2-3 单击"查看页面源代码"选项

在打开的页面中即可查看页面源代码,如图 2-4 所示。

```
          </div>
        </div>
      </div>
      <div class="page-wrapper">
        <div class="container-xl">
          <!-- Page title -->
          <div class="page-header d-print-none">
            <div class="row align-items-center">
              <div class="col">
                <h2 class="page-title">
                  Tables
                </h2>
              </div>
            </div>
          </div>
        </div>
        <div class="page-body">
          <div class="container-xl">
            <div class="row row-cards">
              <div class="col-12">
                <div class="card">
                  <div class="table-responsive">
                    <table
        class="table table-vcenter card-table">
                      <thead>
                        <tr>
                          <th>Name</th>
                          <th>Title</th>
                          <th>Email</th>
                          <th>Role</th>
                          <th class="w-1"></th>
                        </tr>
                      </thead>
                      <tbody>
                        <tr>
                          <td >Paweł Kuna</td>
                          <td class="text-muted" >
```

图 2-4　查看页面源代码

全选源代码，复制这些代码至本地设备，并将其保存为名为“report.html”的文件。

第 3 步，通过浏览器打开刚刚保存的 report.html 文件，如图 2-5 所示。

图 2-5　打开 report.html 文件

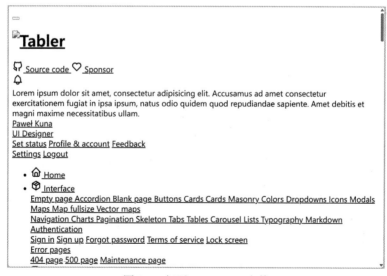

此时，因为在本地的 HTML 页面中缺失引用样式，所以看到的页面是混乱的。重新回到主题页面，通过前端开发者工具查看页面中引用的 JavaScript 和 CSS 文件地址，如图 2-6 所示。

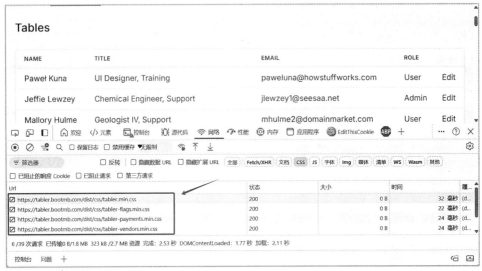

图 2-6　查看页面中引用的 JavaScript 和 CSS 文件地址

将这些地址替换成 report.html 文件中的地址。

比如，将下面的地址

```
<!-- CSS/JS files -->
<link href="./dist/css/tabler.min.css" rel="stylesheet"/>
<link href="./dist/css/tabler-flags.min.css" rel="stylesheet"/>
<link href="./dist/css/tabler-payments.min.css" rel="stylesheet"/>
<link href="./dist/css/tabler-vendors.min.css" rel="stylesheet"/>
<script src="./dist/js/tabler.min.js"></script>
```

替换为

```
<!-- CSS/JS files -->
<link href="https://www.example.com/dist/css/tabler.min.css" rel="stylesheet"/>
<link href="https://www.example.com/dist/css/tabler-flags.min.css" rel="stylesheet"/>
<link href="https://www.examle.com/dist/css/tabler-payments.min.css" rel="stylesheet"/>
<link href="https://www.example.com/dist/css/tabler-vendors.min.css" rel="stylesheet"/>
<script src="https://www.example.com/dist/js/tabler.min.js"></script>
```

刷新本地的 report.html 页面，就可以正常显示了，如图 2-7 所示。

接下来对 report.html 页面的内容进行修改，保留我们需要的功能。当然，这需要你具备一定的 HTML 和 CSS 基础知识。

第 2 章　自动化测试报告设计　37

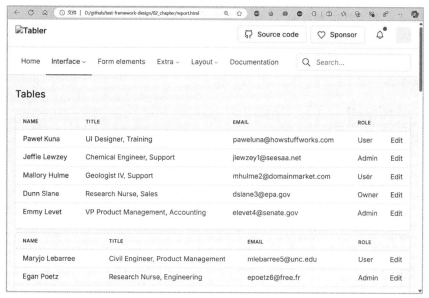

图 2-7　本地的 report.html 页面

2.3　Jinja 模板引擎

Jinja 是一个表达能力强、可扩展的模板引擎。模板中的特殊占位符允许我们编写类似于 Python 语法的代码，之后即可向模板传递数据以呈现最终文档。

用 pip 命令安装 Jinja2。

```
> pip install Jinja2
```

2.3.1　Jinja2 的基础使用方法

（1）Jinja2 中的模板变量如下。

```
from jinja2 import Template


template = Template('Hello {{ name }}!')
tmp = template.render(name='John Doe')
print(tmp)
```

hello {{ name }} 是定义的模板，{{ name }} 是定义的变量。template.render(name='John Doe') 可对模板中的变量 name 赋值。这其实类似于 Python 中的字符串格式化。

⌛ 运行结果

```
> python demo_01.py
hello John Doe!
```

（2）Jinja2 中的 HTML 模板如下。

```python
from jinja2 import Template

template = Template("""
<html>
<body>
<h1> bobby: </h1>
{% for item in hobby %}
    <input type="checkbox" checked>{{item}}
{% endfor %}
</body>
</html>
""")

tmp = template.render(hobby=["eat", "sleep", "code"])
print(tmp)
```

📖 **代码说明**

模板中定义了 HTML 标签。"{% for i in xxx %} ...{% endfor %}"是模板语言 for 循环的写法，"hobby"用于传入列表。

⏳ **运行结果**

```
> python demo_02.py

<html>
<body>
<h1> bobby: </h1>
    <input type="checkbox" checked>eat
    <input type="checkbox" checked>sleep
    <input type="checkbox" checked>code
</body>
</html>
```

输出的是一段标准的 HTML 代码，先将这段 HTML 代码保存为 demo.html 文件，再通过浏览器打开，即可看到页面中有三个复选框。

2.3.2 在 Python 中使用 Jinja2

当 HTML 和 Python 代码混合在一起使用时非常不方便维护，利用 Jinja2 可以将 HTML 文件导入 Python 中使用。接下来，我们通过一个简单的例子将 Python 中的数据写入 HTML 表格。

📁 **目录结构**

```
02_chapter
├──jinaja2_demo/
│   ├──templates
```

```
|   └──table.html
|   ├──jinja2_demo.py
|   └──result.html
```

创建 table.html 文件,实现一个表格。

```html
<html>
<body>
<h1> case list: </h1>
<table border="1">
    <tr>
        <th>name</th>
        <th>doc</th>
    </tr>

    {% for item in cases %}
    <tr>
        <td>{{ item.name }}</td>
        <td>{{ item.doc }}</td>
    </tr>
    {% endfor %}

</table>
</body>
</html>
```

引入 HTML 模板。

```python
from jinja2 import Environment, FileSystemLoader, select_autoescape

# 加载 templates 目录
env = Environment(
    loader=FileSystemLoader("templates"),
    autoescape=select_autoescape())

# 指定 table.html 文件
template = env.get_template("table.html")

# 准备写入数据
cases_list = [
    {"name": "test_pass", "doc": "pass case"},
    {"name": "test_skip", "doc": "skip case"},
    {"name": "test_fail", "doc": "fail case"},
    {"name": "test_error", "doc": "error case"}]

tmp = template.render(cases=cases_list)
# 保存 HTML 结果
with open("./result.html", "w", encoding="utf-8") as f:
    f.write(tmp)
```

代码说明

整个过程可以分为以下三步。

（1）引入 HTML 模板。

（2）将数据写入（填充）模板。

（3）将模板保存为 HTML 文件。

最终得到 result.html 文件，内容如下。

```html
<html>
<body>
<h1> case list: </h1>
<table border="1">
    <tr>
        <th>name</th>
        <th>doc</th>
    </tr>
    <tr>
        <td>test_pass</td>
        <td>pass case</td>
    </tr>
    <tr>
        <td>test_skip</td>
        <td>skip case</td>
    </tr>
    <tr>
        <td>test_fail</td>
        <td>fail case</td>
    </tr>
    <tr>
        <td>test_error</td>
        <td>error case</td>
    </tr>
</table>
</body>
</html>
```

2.4 基于 unittest 生成 HTML 测试报告

本节会对前面三节所学的知识进行综合运用，即基于 unittest 生成 HTML 测试报告。

目录结构

```
02_chapter
├──jsonrunner
│   ├──result.py
```

```
│   ├── runner.py
│   └── template
│       └── report.html
├── test_htmlrunner.py
├── result.html
└── result.json
```

2.4.1 修改 runner.py 测试运行器

在 2.1 节中，我们在 runner.py 文件中实现了对测试结果的收集，并生成了 JSON 格式的报告，本节我们会在 runner.py 文件的基础上修改代码，使其支持 HTML 测试报告，新增内容如下。

```python
import os
import json
import datetime
from .result import _TestResult
from jinja2 import Environment, FileSystemLoader, select_autoescape

# 指定template目录的路径
PATH_DIR = os.path.dirname(os.path.abspath(__file__))
TEMP_DIR = os.path.join(PATH_DIR, "template")

# 加载template目录
env = Environment(
    loader=FileSystemLoader(TEMP_DIR),
    autoescape=select_autoescape())

# 指定table.html文件
template = env.get_template("report.html")

# 定义测试用例的类型
case_type = {
    0: "passed",
    1: "failure",
    2: "errors",
    3: "skipped"
}

class JSONTestRunner:
    """
    运行测试：生成JSON格式的报告
    """

    def __init__(self, output, verbosity=1):
        self.output = output
        self.verbosity = verbosity
        self.start_time = datetime.datetime.now()
```

```python
    def run(self, test):
        """
        运行测试
        """
        result = _TestResult(self.verbosity)
        test(result)
        case_info = self.test_result(result)

        # 将测试结果转为测试报告
        self.result_to_report(case_info)

        stop_time = datetime.datetime.now()
        print(f"Time Elapsed: {stop_time - self.start_time}")
        return result

    def test_result(self, result):
        ...

    @staticmethod
    def sort_result(result_list):
        ...

    @staticmethod
    def generate_case_data(cid, tid, n, t, e):
        ...

    def result_to_report(self, result):
        """
        将测试结果转为测试报告
        """
        # 保存为 JSON 文件
        with open(self.output, "w", encoding="utf-8") as json_file:
            json.dump(result, json_file, ensure_ascii=False, indent=2)

        # 替换后缀，保存为 HTML 文件
        html_file_name = self.output.replace(".json", ".html")
        tmp = template.render(class_list=result)
        with open(html_file_name, "w", encoding="utf-8") as f:
            f.write(tmp)
```

📖 代码说明

注：使用"..."的地方，表示与 2.1 节的内容一致，无任何改动。

首先，根据目录结构，指定 template 目录的路径，导入 Jinja2 模板，并指定 report.html 文件。

其次，实现 result_to_report()方法，将测试结果数据传给 HTML 模板，根据传入的 self.output

文件名，修改后缀名，并保存为×××.html 文件。

最后，在 run()方法中增加 result_to_report()方法调用，将测试结果分别保存×××.json 文件和×××.html 文件。

2.4.2　修改 HTML 模板

2.2 节介绍了如何获取 HTML 主题模板，下面我们在获取的 HTML 模板中加入 Jinja2 模板语言。

```
<!--report.html-->
...
<div class="card">
  {% for class in class_list %}
  <div class="table-responsive">
    <table class="table table-vcenter card-table">
      <thead>
      <tr>
        <th class="p-0 fw-bolder min-w-100px">{{ class.name }}</th>
        <th >{{ class.desc }}</th>
        <th >总数：{{ class.count }}</th>
            <th></th>
      <th>通过：{{ class.pass }} 失败：{{ class.fail }} 错误：{{ class.error }} 跳过：{{ class.skipped }}
 </th>
        </tr>
        </thead>
        <tbody>
      {% for case in class.cases %}
        <tr>
      <td class="text-muted">{{ case.number }}
</td>
        <td>{{ case.name }}</td>
        <td>{{ case.doc }}</td>
        <td>{{ case.result }}</td>
        <td>{{ case.error }}</td>
        </tr>
      {% endfor %}
       </tbody>
      </table>
     </div>
      {% endfor %}
</div>
...
```

代码说明

由于 HTML 代码比较长,所以上面提供的不是完整的 HTML 模板代码。我们可以在模板中加入更多内容,比如,报告的标题、测试人员,以及运行环境等。

利用 Jinja2 模板语言的两个 for 循环遍历测试数据,并将其插入对应的 HTML 标签中。

另外,页面呈现的样式主要与 HTML 中代码元素的 class 属性和引入的 style.css 样式库有关。比如, class="table-responsive" 具体呈现的样式取决于 CSS 中 table-responsive 类定义的属性。

2.4.3 编写测试用例

编写测试用例,验证修改后的 jsonrunner 是否能生成 HTML 测试报告。

```python
import unittest
from jsonrunner.runner import JSONTestRunner

class TestDemo(unittest.TestCase):
    """测试 Test Demo 类"""

    def test_pass(self):
        """成功"""
        self.assertEqual(5, 5)

    def test_fail(self):
        """失败"""
        self.assertEqual(5, 6)

    def test_error(self):
        """错误"""
        self.assertEqual(a, 6)

    @unittest.skip("skip case")
    def test_skip(self):
        """跳过"""
        ...


if __name__ == '__main__':
    suit = unittest.TestSuite()
    suit.addTests([
        TestDemo("test_pass"),
        TestDemo("test_skip"),
        TestDemo("test_fail"),
        TestDemo("test_error")
    ])
```

```
runner = JSONTestRunner(output="./result.json")
runner.run(suit)
```

测试用例与 2.1 节一致，当然，你也可以尝试增加一些新的测试用例。

⏳ 运行结果

```
> python test_jsonrunner.py
```

运行完成，生成 result.json 和 result.html 文件，其中 result.html 文件如图 2-8 所示。

图 2-8　result.html 文件

第 3 章 数据驱动设计

数据驱动是自动化测试框架非常重要的功能之一,通过对测试数据的管理,可以大大地节省编写测试用例的代码量,尤其是在做接口自动化测试时,由于其自身的特点,数据驱动可以节省很多样例代码,甚至可以使用数据文件管理测试用例。

3.1 unittest 的数据驱动扩展

unittest 提供了两个扩展来实现数据驱动。

3.1.1 parameterized 库

在 Python 单元测试框架中,参数化测试表现并不理想,为了解决这个问题,parameterized 项目应运而生。

parameterized 支持的单元测试框架如图 3-1 所示。

	Py3.7	Py3.8	Py3.9	Py3.10	Py3.11	PyPy3	@mock.patch
nose	yes	yes	yes	yes	no§	no§	yes
nose2	yes	yes	yes	yes	yes	yes	yes
py.test 2	no*	no*	no*	no*	no*	no*	no*
py.test 3	yes	yes	yes	yes	no*	no*	yes
py.test 4	no**	no**	no**	no**	no**	no**	no**
py.test fixtures	no†	no†	no†	no†	no†	no†	no†
unittest (@parameterized.expand)	yes	yes	yes	yes	yes	yes	yes
unittest2 (@parameterized.expand)	yes	yes	yes	no§	no§	yes	

图 3-1 parameterized 支持的单元测试框架

从图 3-1 中可以看出，parameterized 仅支持 py.test 3，nose 已经不再维护，nose2 的定位是 unittest 的扩展，因此，parameterized 目前主要作为 unittest 的扩展被使用。

用 pip 命令安装 parameterized 库。

```
> pip install parameterized
```

使用示例

通过 parameterized 实现参数化。

```python
# parameterized_demo.py
import unittest
from parameterized import parameterized

class AddTestCase(unittest.TestCase):

    @parameterized.expand([
        ("2 and 3", 2, 3, 5),
        ("10 and 20", 10, 20, 30),
        ("hello and word", "hello", "world", "helloworld"),
    ])
    def test_add(self, _, a, b, expected):
        self.assertEqual(a + b, expected)

if __name__ == '__main__':
    unittest.main(verbosity=2)
```

代码说明

首先，导入 parameterized 库下面的 parameterized 类。

其次，通过 parameterized.expand 装饰器装饰 test_add() 测试用例。

在 parameterized.expand 装饰器中，每个元组都被视作一个单独的测试用例。parameterized 默认为每个测试用例都添加了数字编号，并把元组的第一个参数解析为测试用例名称的后缀。

最后，使用 unittest 的 main() 方法，设置 verbosity 参数为 2，以输出更详细的运行日志。

运行结果

```
> python parameterized_demo.py

test_add_0_2_and_3 (__main__.AddTestCase) ... ok
test_add_1_10_and_20 (__main__.AddTestCase) ... ok
test_add_2_hello_and_word (__main__.AddTestCase) ... ok

----------------------------------------------------------------------
Ran 3 tests in 0.001s

OK
```

从运行结果可以看出，测试用例的名称=测试方法名+编号+第一个参数，即 test_add + 0 + 2_and_3。这样的取值规则存在缺陷，如果元组的第一个参数是中文的，则无法正常显示。

3.1.2 DDT 库

DDT（Data-Driven Tests）是针对 unittest 框架设计的扩展库。它允许使用不同的测试数据来运行一个测试用例，并将其展示为多个测试用例。

用 pip 命令安装 DDT 库。

```
> pip install ddt
> pip install pyyaml
```

注：DDT 支持 YAML 格式的数据驱动文件，前提是需要安装 pyyaml 库。

使用示例

通过 DDT 实现参数化。

```python
# ddt_demo.py
import unittest
from ddt import ddt, data, unpack

@ddt
class TestDTT(unittest.TestCase):

    @data([1, 2, 3], [4, 5, 9], [6, 7, 13])
    @unpack
    def test_add_list(self, a, b, c):
        self.assertEqual(a+b, c)

    @data(("Hi", "HI"), ("hello", "HELLO"), ("world", "WORLD"))
    @unpack
    def test_upper_tuple(self, s1, s2):
        self.assertEqual(s1.upper(), s2)

    @data({"d": "hello world", "t": str},
          {"d": 110, "t": int},
          {"d": True, "t": bool})
    @unpack
    def test_data_type_dict(self, d, t):
        self.assertTrue(isinstance(d, t))

if __name__ == '__main__':
    unittest.main(verbosity=2)
```

代码说明

测试类需要通过 ddt 装饰器进行装饰。

DDT 提供了不同形式的参数化。这里列举了三组参数化：第一组为列表（list），第二组为元组（tuple），第三组为字典（dict）。

需要注意的是，字典的 key 与测试方法的参数必须同名。

运行结果

```
> python ddt_demo.py
test_add_list_1__1__2__3_ (__main__.TestBaidu) ... ok
test_add_list_2__4__5__9_ (__main__.TestBaidu) ... ok
test_add_list_3__6__7__13_ (__main__.TestBaidu) ... ok
test_upper_tuple_1__Hi___HI__ (__main__.TestBaidu) ... ok
test_upper_tuple_2__hello___HELLO__ (__main__.TestBaidu) ... ok
test_upper_tuple_3__world___WORLD__ (__main__.TestBaidu) ... ok
test_data_type_dict_1 (__main__.TestBaidu) ... ok
test_data_type_dict_2 (__main__.TestBaidu) ... ok
test_data_type_dict_3 (__main__.TestBaidu) ... ok

----------------------------------------------------------------------
Ran 9 tests in 0.002s

OK
```

从运行结果可以看出，如果是列表（list）和元组（tuple），那么测试用例的名称=测试方法名+全部参数；如果是字典（dict），那么测试用例的名称=测试方法名+数字编号。

使用示例

DDT 的数据文件用法是利用 JSON 或 YAML 文件实现参数化。

目录结构

```
03_chapter/
├──libs_demo
│   ├──test_data    # 数据驱动文件目录
│   │   ├──test_data.json
│   │   └──test_data.yaml
│   └──ddt_demo2.py
```

创建 test_data.json 文件，测试数据如下。

```
{
    "positive_integer_range": {
        "start": 0,
        "end": 2,
        "value": 1
    },
    "negative_integer_range": {
        "start": -2,
        "end": 0,
```

```
        "value": -1
    },
    "positive_real_range": {
        "start": 0.0,
        "end": 1.0,
        "value": 0.5
    }
}
```

创建 test_data.yaml 文件，测试数据如下。

```
positive_integer_range:
    start: 0
    end: 2
    value: 1
negative_integer_range:
    start: -2
    end: 0
    value: -1
positive_real_range:
    start: 0.0
    end: 1.0
    value: 0.5
```

编写测试用例，引用 JSON 或 YAML 文件。

```
# ddt_demo2.py
import unittest
from ddt import ddt, file_data

@ddt
class TestDDT(unittest.TestCase):

    @file_data('test_data/test_data.json')
    def test_file_data_json(self, start, end, value):
        self.assertLess(start, end)
        self.assertLess(value, end)
        self.assertGreater(value, start)

    @file_data('test_data/test_data.yaml')
    def test_file_data_yaml(self, start, end, value):
        self.assertLess(start, end)
        self.assertLess(value, end)
        self.assertGreater(value, start)

if __name__ == '__main__':
    unittest.main(verbosity=2)
```

📖 代码说明

首先，将测试数据写入 JSON 或 YAML 文件。然后，通过 file_data 装饰器指定测试文件的路径，这里的路径为相对路径。

注意，如果需要使用 YAML 作为数据文件，则需要安装 pyyaml 库。

运行结果

```
> python ddt_demo2.py

test_file_data_json_1_positive_integer_range (__main__.TestDDT)
test_file_data_json_1_positive_integer_range ... ok
test_file_data_json_2_negative_integer_range (__main__.TestDDT)
test_file_data_json_2_negative_integer_range ... ok
test_file_data_json_3_positive_real_range (__main__.TestDDT)
test_file_data_json_3_positive_real_range ... ok
test_file_data_yaml_1_positive_integer_range (__main__.TestDDT)
test_file_data_yaml_1_positive_integer_range ... ok
test_file_data_yaml_2_negative_integer_range (__main__.TestDDT)
test_file_data_yaml_2_negative_integer_range ... ok
test_file_data_yaml_3_positive_real_range (__main__.TestDDT)
test_file_data_yaml_3_positive_real_range ... ok

----------------------------------------------------------------------
Ran 6 tests in 0.002s

OK
```

3.2 参数化装饰器

parameterized 库和 DDT 库作为 unittest 的数据驱动扩展库，在使用过程中存在以下不足之处。

■ parameterized 库的不足。

首先，parameterized 库不支持测试数据文件。通过 parameterized.expand 装饰器在代码中定义数据，仅适合测试数据较少的情况；对于复杂或数据量较大的测试数据（比如，我们在做接口测试时，测试数据会多达数百条），在代码中管理则显然既不够优雅，也不方便维护。

其次，parameterized.expand 装饰器不支持 dict 格式的数据，而 dict 格式可以更好地描述测试数据。

■ DDT 库的不足。

首先，DDT 库在使用过程中需要用到多个装饰器。ddt 装饰器用于装饰测试类，data 装饰器用于设置测试数据，unpack 装饰器用于解压缩数据，这些装饰器看上去有点儿冗余。

其次，虽然 DDT 库支持 JSON 和 YAML 格式的数据文件，但是，不支持 CSV 和 xlsx 等常见格式的数据文件。

基于上面的不足，我们设计封装自己的数据驱动库。

📁 **目录结构**

```
03_chapter/
├──extend/
│   └──parameterized_extend.py
└──parameterized_extend_demo.py
```

3.2.1 实现 data 装饰器

parameterized 库的核心代码大约有 600 行，建议花一些时间阅读这部分代码，这对于我们深入理解数据驱动的实现很有帮助。我们可以在 parameterized 库的基础上进行开发。

在使用 parameterized 库时，可以感受到 parameterized 库对 unittest 的支持是个"二等公民"，因为其装饰器的写法是 parameterized.expand()，表明它是通过一个 expand()方法实现的。

创建 parameterized_extend.py 文件。

```python
# extend/parameterized_extend.py
import warnings
from parameterized.parameterized import inspect
from parameterized.parameterized import parameterized
from parameterized.parameterized import skip_on_empty_helper
from parameterized.parameterized import reapply_patches_if_need
from parameterized.parameterized import delete_patches_if_need
from parameterized.parameterized import default_doc_func
from parameterized.parameterized import default_name_func
from parameterized.parameterized import wraps


def data(input, name_func=None, doc_func=None, skip_on_empty=False, **legacy):
    """
    重写 parameterized.expand()方法
    """

    if "testcase_func_name" in legacy:
        warnings.warn("testcase_func_name= is deprecated; use name_func=",
                      DeprecationWarning, stacklevel=2)
        if not name_func:
            name_func = legacy["testcase_func_name"]

    if "testcase_func_doc" in legacy:
        warnings.warn("testcase_func_doc= is deprecated; use doc_func=",
                      DeprecationWarning, stacklevel=2)
        if not doc_func:
            doc_func = legacy["testcase_func_doc"]

    doc_func = doc_func or default_doc_func
```

```python
        name_func = name_func or default_name_func

        def parameterized_expand_wrapper(f, instance=None):
            frame_locals = inspect.currentframe().f_back.f_locals

            parameters = parameterized.input_as_callable(input)()

            if not parameters:
                if not skip_on_empty:
                    raise ValueError(
                        "Parameters iterable is empty (hint: use "
                        "`parameterized.expand([], skip_on_empty=True)` to skip "
                        "this test when the input is empty)"
                    )
                return wraps(f)(skip_on_empty_helper)

            digits = len(str(len(parameters) - 1))
            for num, p in enumerate(parameters):
                name = name_func(f, "{num:0>{digits}}".format(digits=digits, num=num), p)
                # If the original function has patches applied by 'mock.patch',
                # re-construct all patches on the just former decoration layer
                # of param_as_standalone_func so as not to share
                # patch objects between new functions
                nf = reapply_patches_if_need(f)
                frame_locals[name] = parameterized.param_as_standalone_func(p, nf, name)
                frame_locals[name].__doc__ = doc_func(f, num, p)

            # Delete original patches to prevent new function from evaluating
            # original patching object as well as re-constructed patches.
            delete_patches_if_need(f)

            f.__test__ = False

        return parameterized_expand_wrapper
```

📖 代码说明

首先，把 parameterized 类下面的 expand() 方法拷贝出来，并将它重命名为 data() 函数。然后，从 parameterized 库的相关文件中导入 data() 函数依赖的方法。

🔔 使用示例

编写测试用例，验证 data() 函数是否可用。

```python
# parameterized_extend_demo.py
import unittest
from extend.parameterized_extend import data

class ParamTest(unittest.TestCase):
```

```python
    @data([
        ("2 and 3", 2, 3, 5),
        ("10 and 20", 10, 20, 30),
        ("hello and word", "hello", "world", "helloworld"),
    ])
    def test_tuple_add(self, _, a, b, expected):
        self.assertEqual(a + b, expected)

    @data([
        ["2 and 3", 2, 3, 5],
        ["10 and 20", 10, 20, 30],
        ["hello and word", "hello", "world", "helloworld"],
    ])
    def test_list_add(self, _, a, b, expected):
        self.assertEqual(a + b, expected)

if __name__ == '__main__':
    unittest.main(verbosity=2)
```

与 parameterized.expand 装饰器的用法相同，只不过将装饰器的名称替换为了 data。现在升级为"一等公民"了。

⌛ 运行结果

```
> python parameterized_extend_demo.py
test_list_add_0_2_and_3 (__main__.ParamTest.test_list_add_0_2_and_3) ... ok
test_list_add_1_10_and_20 (__main__.ParamTest.test_list_add_1_10_and_20) ... ok
test_list_add_2_hello_and_word
(__main__.ParamTest.test_list_add_2_hello_and_word) ... ok
test_tuple_add_0_2_and_3 (__main__.ParamTest.test_tuple_add_0_2_and_3) ... ok
test_tuple_add_1_10_and_20 (__main__.ParamTest.test_tuple_add_1_10_and_20) ... ok
test_tuple_add_2_hello_and_word
(__main__.ParamTest.test_tuple_add_2_hello_and_word) ... ok
----------------------------------------------------------------------
Ran 6 tests in 0.001s

OK
```

3.2.2 实现 dict 数据格式支持

目前 data 装饰器仅支持列表（list）和元组（tuple）两种数据格式，虽然这两种数据格式比较简捷，但是数据的可读性较差。

```python
@data([
    ("case1", "", "123"),
    ("case2", "user", ""),
    ("case3", "user", "123")])
```

从上面的例子可以看出，如果单看元组（tuple）中的每列数据，我们很难理解它们所代表的含义。如果把它转换成字典（dict）的 key 来定义每个字段的含义，那么理解起来会更加容易。

```python
@data([
    {"scene": "username_is_null", "username": "", "password": "123"},
    {"scene": "password_is_null", "username": "user", "password": ""},
    {"scene": "login_success", "username": "user", "password": "123"},])
```

现在需求有了，我们要做的就是修改 data() 函数，让它可以支持字典（list）格式的数据。

```python
...
def check_data(list_data: list) -> list:
    """
    检查数据格式，如果是 dict, 就将其转化为 list
    """
    if isinstance(list_data, list) is False:
        raise TypeError("The data format is not `list`.")
    if len(list_data) == 0:
        raise ValueError("The data format cannot be `[]`.")
    if isinstance(list_data[0], dict):
        test_data = []
        for data_ in list_data:
            line = []
            for d in data_.values():
                line.append(d)
            test_data.append(line)
        return test_data

    return list_data


def data(input, name_func=None, doc_func=None, skip_on_empty=False, **legacy):
    """
    重写 parameterized.expand() 方法
    """
    input = check_data(input)
....
```

📖 **代码说明**

实现 check_data() 函数，首先，进行数据格式检查。如果判断传入的数据类型为字典（dict），则通过 values() 方法取出字典中的每个值（value），将它们组合成一个二维列表，并返回该列表。

然后，在 data() 函数中，调用 check_data() 函数进行判断。

🔖 **使用示例**

编写测试用例，验证上面的改动是否有效。

```python
import unittest
from extend.parameterized_extend import data


class ParamTestCase(unittest.TestCase):

    @data([
        {"scene": "username_is_null", "username": "", "password": "123"},
        {"scene": "password_is_null", "username": "user", "password": ""},
        {"scene": "login_success", "username": "user", "password": "123"},
    ])
    def test_dict_data(self, scene, username, password):
        ...


if __name__ == '__main__':
    unittest.main(verbosity=2)
```

⏳ 运行结果

```
> python parameterized_extend_demo.py
test_dict_data_0_username_is_null
(__main__.ParamTest.test_dict_data_0_username_is_null) ... ok
test_dict_data_1_password_is_null
(__main__.ParamTest.test_dict_data_1_password_is_null) ... ok
test_dict_data_2_login_success
(__main__.ParamTest.test_dict_data_2_login_success) ... ok

----------------------------------------------------------------------
Ran 3 tests in 0.001s

OK
```

3.3 参数化数据文件

当管理大量测试数据时，无疑需要用到测试文件，常见的测试文件格式有 JSON、YAML、Excel 和 CSV 等。

首先，我们要确认数据驱动能够支持数据文件，因此需要单独设计一个装饰器来支持数据文件。注意，不同的数据文件，读取方式也不同。

其次，我们要思考如何将数据文件中的数据用于数据驱动。方法很简单，不论哪种类型的数据文件，只需将其数据最终转化为 data() 函数所需的数据格式，由该函数处理即可。

📂 目录结构

```
03_chapter/
├──test_data/
```

```
    │   ├──json_data.json
    │   ├──yaml_data.yaml
    │   ├──csv_data.csv
    │   └──excel_data.xlsx
    ├──extend/
    │   ├──parameterized_extend.py
    │   └──data_conversion.py
    └──parameterized_extend_demo.py
```

3.3.1 JSON 数据转换

JSON 是存储测试数据的常见格式。下面我们读取 JSON 文件,将文件中的数据转换为 list 格式。

创建 data_conversion.py 文件,实现 json_to_list()函数。

```
import json
from extend.parameterized_extend import check_data


def json_to_list(file: str = None, key: str = None) -> list:
    """
    将 JSON 文件中的数据转换为 list 格式
    :param file: JSON 文件的路径
    :param key: 字典的 key,指定解析哪部分数据
    return: list
    """
    if file is None:
        raise FileExistsError("Please specify the JSON file to convert.")

    if key is None:
        with open(file, "r", encoding="utf-8") as json_file:
            data = json.load(json_file)
            list_data = check_data(data)
    else:
        with open(file, "r", encoding="utf-8") as json_file:
            try:
                data = json.load(json_file)[key]
                list_data = check_data(data)
            except KeyError:
                raise ValueError(f"Check the test data, no '{key}'.")

    return list_data
```

📖 代码说明

创建 json_to_list()函数,使其接收两个参数:file 用于指定 JSON 文件的路径;key 用于指定解析哪部分数据。如果不指定 key,则默认解析整个文件。

代码实现部分比较简单。首先,通过 open()方法打开一个 JSON 文件,利用 json 模块提供的 load() 方法将读取的 JSON 文件中的数据转换为 Python 字典格式。其次,调用 check_data() 方法判断数据格式。如果是 dict 格式,则将其转换为 list 格式,以保证返回的数据格式统一为 list。

使用示例

下面编写测试用例,验证 json_to_list()函数是否可用。在 test_data 目录中创建 json_data.json 测试文件,内容如下。

```
{
    "login": [
        {"scene": "username_is_null", "username": "", "password": "123"},
        {"scene": "password_is_null", "username": "user", "password": ""},
        {"scene": "login_success", "username": "user", "password": "123"}
    ]
}
```

创建测试用例。

```
import unittest
from extend.parameterized_extend import data
from extend.data_conversion import json_to_list

class FileParamTest(unittest.TestCase):

    @data(json_to_list("test_data/json_data.json", key="login"))
    def test_login_json(self, name, username, password):
        print(f"scene: {name}, username: '{username}' password: '{password}'")


if __name__ == '__main__':
    unittest.main()
```

在 data 装饰器中嵌套使用 json_to_list() 函数显得不够优雅,后续我们将采用 file_data 装饰器统一处理数据文件。

运行结果

```
> python parameterized_extend_demo.py
scene: username_is_null, username: '' password: '123'
.scene: password_is_null, username: 'user' password: ''
.scene: login_success, username: 'user' password: '123'
.
----------------------------------------------------------------------
Ran 3 tests in 0.001s

OK
```

3.3.2 YAML 数据转换

YAML 文件与 JSON 文件可以相互转换。与 JSON 文件相比，YAML 文件在数据的描述上更加简捷。在 Python 的标准库中并没有提供对 YAML 文件的支持，因此我们需要使用 pyyaml 库。

用 pip 命令安装 pyyaml 库。

```
> pip install pyyaml
```

读取 YAML 文件中的数据并将其转换为 list 格式，在 data_conversion.py 文件中实现 yaml_to_list()函数。

```python
import yaml

def yaml_to_list(file: str = None, key: str = None) -> list:
    """
    将 YAML 文件中的数据转换为 list 格式
    :param file: YAML 文件的路径
    :param key: 字典的 key，指定解析哪部分数据
    return: list
    """
    if file is None:
        raise FileExistsError("Please specify the YAML file to convert.")

    if key is None:
        with open(file, "r", encoding="utf-8") as yaml_file:
            data = yaml.load(yaml_file, Loader=yaml.FullLoader)
            list_data = check_data(data)
    else:
        with open(file, "r", encoding="utf-8") as yaml_file:
            try:
                data = yaml.load(yaml_file, Loader=yaml.FullLoader)[key]
                list_data = check_data(data)
            except KeyError as exc:
                raise ValueError(f"Check the YAML test data, no '{key}'") from exc

    return list_data
```

📖 代码说明

创建 yaml_to_list()函数，使其接收两个参数：file 用于指定 YAML 文件的路径；key 用于指定解析哪部分数据。如果不指定 key，则默认解析整个文件。

代码实现部分比较简单。首先，通过 open() 方法打开一个 YAML 文件，利用 YAML 文件提供的 load()方法将读取的 YAML 文件中的数据转换为 Python 字典格式。其次，调用 check_data()方法判断数据格式。如果是 dict 格式，则将其转换为 list 格式，以保证返回的数据格式统一为 list。

使用示例

下面编写测试用例，验证 yaml_to_list() 函数是否可用。在 test_data 目录中创建 yaml_data.yaml 数据文件。

```yaml
---
login:
- scene: username_is_null
  username: ''
  password: '123'
- scene: password_is_null
  username: user
  password: ''
- scene: login_success
  username: user
  password: '123'
```

如果不会创建 YAML 文件，那么可以借助在线工具，将 JSON 文件转换为 YAML 文件。

创建测试用例。

```python
import unittest
from extend.parameterized_extend import data
from extend.data_conversion import yaml_to_list


class FileParamTest(unittest.TestCase):

    @data(yaml_to_list("test_data/yaml_data.yaml", key="login"))
    def test_login_yaml(self, name, username, password):
        print(f"scene: {name}, username: '{username}' password: '{password}'")

if __name__ == '__main__':
    unittest.main()
```

从上面的代码中可以看出，yaml_to_list() 函数的用法与 json_to_list() 函数的用法基本一致。

3.3.3 CSV 数据转换

CSV 文件是比较常用的一种数据格式。在 Python 标准库中，提供了操作 CSV 文件的相关功能。

读取 CSV 文件中的数据并将其转换为 list 格式，在 data_conversion.py 文件中实现 csv_to_list() 函数。

```python
import csv
import codecs
from itertools import islice


def csv_to_list(file: str = None, line: int = 1) -> list:
    """
```

```
将 CSV 文件转换为 list 格式
:param file: CSV 文件的路径
:param line: 从第几行开始读取数据
:return: list
"""
if file is None:
    raise FileExistsError("Please specify the CSV file to convert.")

table_data = []
with codecs.open(file, 'r', encoding='utf_8_sig') as csv_file:
    csv_data = csv.reader(csv_file)
    for i in islice(csv_data, line - 1, None):
        table_data.append(i)

return table_data
```

📖 **代码说明**

创建 csv_to_list() 函数，使其接收两个参数：file 用于指定 CSV 文件的路径；line 指定从第几行开始读取数据。CSV 文件与 Excel 文件类似，数据都是以表格的形式存储的。因为我们习惯将第 1 行作为表头来定义每列数据的名称，所以可以指定从第 2 行开始读取数据。

在读取文件部分，首先，通过 codecs.open() 方法打开一个 CSV 文件。然后，通过 csv.reader() 函数读取 csv_file 文件对象。接着，通过 islice() 方法对读取的数据进行遍历，将遍历的每行数据都追加到 table_data 列表。最后，返回 table_data 列表。

🔍 **使用示例**

下面编写测试用例，验证 csv_to_list() 函数是否可用。在 test_data 目录中创建 csv_data.csv 文件，CSV 格式的测试数据如图 3-2 所示。

图 3-2　CSV 格式的测试数据

创建测试用例。

```python
import unittest
from extend.parameterized_extend import data
from extend.data_conversion import csv_to_list

class FileParamTest(unittest.TestCase):

    @data(csv_to_list("test_data/csv_data.csv", line=2))
    def test_login_csv(self, name, username, password):
        print(f"scene: {name}, username: '{username}' password: '{password}'")

if __name__ == '__main__':
    unittest.main()
```

注意，csv_to_list() 函数的第二个参数 line 用于指定从第几行开始读取数据。

3.3.4　Excel 数据转换

Excel 文件的应用非常广泛，与 CSV 文件相比，Excel 文件支持多标签页（Sheet）。在 Python 标准库中并没有提供对 Excel 文件的支持，因此我们需要使用第三方库来操作 Excel 文件，如 openpyxl 和 pandas。

我们这里选择用 openpyxl，因为它更加轻量一些，仅提供对 Excel 文件的操作。而 pandas 专注于数据分析，功能更加强大，同时体积也较大。

用 pip 命令安装 openpyxl。

```
> pip install openpyxl
```

读取 Excel 文件中的数据并将其转换为 list 格式，在 data_conversion.py 文件中实现 excel_to_list()函数。

```python
from openpyxl import load_workbook

def excel_to_list(file: str = None, sheet: str = "Sheet1", line: int = 1) -> list:
    """
    将 Excel 文件中的数据转换为 list 格式
    :param file: Excel 文件的路径
    :param sheet: 标签页的名称，默认为 Sheet1
    :param line: 从第几行开始读取数据
    :return: list data
    """
    if file is None:
        raise FileExistsError("Please specify the Excel file to convert.")

    excel_table = load_workbook(file)
    sheet = excel_table[sheet]
```

```
    table_data = []
    for i in sheet.iter_rows(line, sheet.max_row):
        line_data = []
        for field in i:
            line_data.append(field.value)
        table_data.append(line_data)

    return table_data
```

📖 **代码说明**

创建 excel_to_list()函数，使其接收三个参数：file 用于指定 Excel 文件的路径；sheet 用于指定标签页的名称，默认为 Sheet1；line 用于指定从第几行开始读取数据。

在读取文件部分，首先，通过 load_workbook() 方法打开一个 Excel 文件，并指定标签页的名称。然后，通过 iter_rows()方法遍历每一行数据，再通过内循环遍历每一列数据，将遍历的数据都保存到 table_data 列表。最后，返回 table_data 列表。

👆 **使用示例**

在 test_data 目录中创建 excel_data.xlsx 文件，Excel 格式的测试数据如图 3-3 所示。

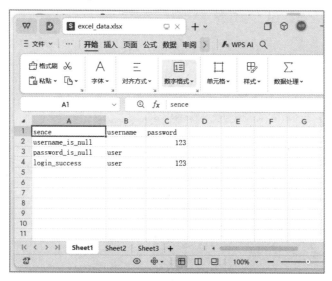

图 3-3　Excel 格式的测试数据

创建测试用例。

```
import unittest
from extend.parameterized_extend import data
from extend.data_conversion import excel_to_list

class FileParamTest(unittest.TestCase):
```

```python
    @data(excel_to_list("test_data/excel_data.xlsx", sheet="Sheet1", line=2))
    def test_login_excel(self, name, username, password):
        print(f"scene: {name}, username: '{username}' password: '{password}'")

if __name__ == '__main__':
    unittest.main()
```

注意，对于 excel_to_list() 函数，除了需要指定文件路径，还需要指定 sheet 和 line 参数。

3.3.5 实现 file_data 装饰器

通过前面的设计，我们分别成功实现了对主流数据文件格式的转换，包括 JSON 文件、YAML 文件、CSV 文件和 Excel 文件。但是，通过 data 装饰器嵌套转换函数这种方式非常不方便，因此我们需要进一步实现 file_data 装饰器，使其可以自动识别文件的类型。

在 data_conversion.py 文件中实现 file_data 装饰器，代码如下。

```python
import os
from extend.parameterized_extend import data

...

def file_data(file: str, line: int = 1, sheet: str = "Sheet1", key: str = None):
    """
    支持文件转参数化
    :param file: 文件名
    :param line: 对于 Excel 文件和 CSV 文件，指定从第几行开始读取数据
    :param sheet: Excel 文件的标签页名称
    :param key: YAML 文件、JSON 文件指定的 key
    """
    if file is None:
        raise FileExistsError("File name does not exist.")

    if os.path.isfile(file) is False:
        raise FileExistsError(f"No '{file}' data file found.")

    suffix = file.split(".")[-1]
    if suffix == "csv":
        data_list = csv_to_list(file, line=line)
    elif suffix == "xlsx":
        data_list = excel_to_list(file, sheet=sheet, line=line)
    elif suffix == "json":
        data_list = json_to_list(file, key=key)
    elif suffix == "yaml":
        data_list = yaml_to_list(file, key=key)
    else:
        raise FileExistsError(f"Your file is not supported: {file}")
```

```
    return data(data_list)
```

📖 **代码说明**

创建 file_data() 函数，使其中的 file 参数可以接收不同类型的文件，而对于其他几个参数，则根据文件类型进行选择性设置。判断文件的后缀名，并交由相应的数据转换函数进行处理，得到 list 格式的数据。

导入 parameterized_extend.py 文件中的 data() 函数，将处理后的 list 格式的数据传给 data() 函数。

💡 **使用示例**

仍然使用前面的数据作为测试数据，将装饰器替换为 file_data，重新实现前面的测试用例。

```python
import unittest
from extend.data_conversion import file_data


class FileDataTest(unittest.TestCase):

    @file_data("test_data/json_data.json", key="login")
    def test_login_json(self, name, username, password):
        print(f"case: {name}, username: '{username}' password: '{password}'")

    @file_data("test_data/yaml_data.yaml", key="login")
    def test_login_yaml(self, name, username, password):
        print(f"case: {name}, username: '{username}' password: '{password}'")

    @file_data("test_data/csv_data.csv", line=2)
    def test_login_csv(self, name, username, password):
        print(f"case: {name}, username: '{username}' password: '{password}'")

    @file_data("test_data/excel_data.xlsx", sheet="Sheet1", line=2)
    def test_login_excel(self, name, username, password):
        print(f"case: {name}, username: '{username}' password: '{password}'")


if __name__ == '__main__':
    unittest.main()
```

相比较而言，file_data 装饰器的用法更加简捷。

这里仍要注意一个问题，即文件的路径和测试用例的执行位置必须满足严格的要求。如果使用相对路径，则需要在测试文件（parameterized_expend_demo.py）的当前目录中运行测试用例；否则需要在指定文件路径时设置绝对路径（比如，d://project/test_data/xxx.json）。

但这两种方式都不够灵活，我们需要进一步实现数据文件的查找规则，让测试用例自动查找距离自己最近的数据驱动文件。这部分内容可以作为一个思考题，由你来思考如何实现。

第 4 章 数据库操作封装设计

在自动化测试过程中，经常涉及对数据库的操作，比如，通过数据库准备测试数据，在数据入库之后对数据进行检查等。基于这个高频的需求，我们可以将对数据库的操作集成到测试框架中。

4.1 操作数据库技术

在 Python 中操作数据库的方式有两种。

数据库驱动：这个是必备的，比如，在操作 SQLite、MySQL、Oracle、MongoDB 和 SQL Server 等数据库时都需要有对应的驱动库。

ORM：Object Relational Mapping 的缩写，中文含义为"对象关系映射"，它解决了对象和关系数据库之间的数据交互问题。与数据库驱动相比，ORM 是更高维度的抽象和封装。

4.1.1 数据库驱动

在编程语言中，连接每一种数据库都需要有对应的驱动。以 MySQL 数据库为例，我们可以使用 PyMySQL 驱动库。

使用 pip 命令安装 PyMySQL 驱动库。

```
> pip install pymysql
```

通过 PyMySQL 驱动库可以连接和操作 MySQL 数据库。

```
# pymysql_demo.py
import pymysql.cursors

# 连接MySQL数据库
```

```python
connection = pymysql.connect(
    host='localhost',        # 数据库 IP 地址
    user='root',             # 用户名
    password='123456',       # 密码
    database='dev_db',       # 数据库名称
    cursorclass=pymysql.cursors.DictCursor
)

with connection as conn:
    with conn.cursor() as cursor:
        # 查询 user 表
        sql = "SELECT `id`, `name`, `email` FROM `user`"
        cursor.execute(sql)
        result = cursor.fetchone()
        print(result)
```

📖 **代码说明**

connect()：用于建立与 MySQL 数据库的连接，需要填写数据库 IP 地址（host）、用户名（user）、密码（password）和数据库名称（database）等。不同数据库的连接方式有所不同。比如，对于 SQLite 数据库，只需指定数据库文件路径即可。

cursor()：建立数据库光标。

execute()：执行 SQL 语句。

fetchone()：查询一条数据，返回 dict 格式的数据。也可以先使用 fetchall() 方法查询所有数据，再返回 list 格式的数据。

⌛ **运行结果**

```
> python pymysql_demo.py

{'id': 1, 'name': 'admin', 'email': 'admin@gmail.com'}
```

4.1.2 ORM

ORM 可以看成是一种基于数据库驱动的高级封装技术。它不直接使用 SQL 语句，而是通过编程语言中创建的对象与数据库表建立映射关系，从而提供数据库操作的 API。这样开发者可以像操作类对象一样操作数据。以 SQLAlchemy 为例。

用 pip 命令安装 SQLAlchemy。

```
> pip install SQLAlchemy
```

实现 user 表的数据查询。

```
# sqlalchemy_demo.py
```

```python
from sqlalchemy import create_engine
from sqlalchemy.orm import declarative_base
from sqlalchemy import Column
from sqlalchemy.types import *
from sqlalchemy.orm import sessionmaker

# 数据库连接
engine = create_engine("mysql+pymysql://root:123456@localhost/dev_db",
connect_args={'charset': 'utf8'})
Session = sessionmaker(bind=engine)
s = Session()

Base = declarative_base()

class User(Base):
    """user 表对象"""

    __tablename__ = 'user'  # 定义数据表名为 user

    # 定义数据库表字段和类型
    id = Column(Integer, primary_key=True)
    name = Column(String(100))
    password = Column(String(100))
    email = Column(String(100))

    def __repr__(self):
        return "<id:%s name:%s email:%s>" % (self.id, self.name, self.email)


# 查询 user 表
user = s.query(User).filter(User.name == 'admin').first()
print(f"id:{user.id}, name:{user.name}, email:{user.email}")
```

📖 **代码说明**

create_engine()：建立 MySQL 数据库连接，需要填写数据库 IP 地址（host）、用户名（user）、密码（password）和数据库名称（database）等。

sessionmaker()：创建 session，绑定数据库引擎。

User 类：User 类继承自 declarative_base()，通过 __tablename__ 定义绑定的数据表名。接下来定义数据库表字段和类型，从而让类对象与数据库表字段产生对应关系。

最后，像操作类对象一样操作数据：通过 query() 方法传入 user 表对象，用 filter() 方法指定查询条件，用 first() 方法返回第一条数据。

⏳ 运行结果

```
> python sqlalchemy_demo.py

id:1, name:admin, email:admin@gmail.com
```

4.2 数据库操作封装

数据库的 SQL 语句大体分为以下两类。

（1）执行 SQL：比如，插入、删除和更新，这类操作一般无须返回数据。

（2）查询 SQL：查询 SQL 一般会返回查询的数据，即单条数据或多条数据。

📁 目录结构

```
04_chapter
├──db_operate/
│   ├──__init__.py
│   └──mysql_db.py
└──test_mysql_operate.py
```

4.2.1 封装执行和查询

下面实现 MySQL 语句的执行和查询方法。在 db_operate/目录中创建 mysql_db.py 文件。

```python
from typing import Any
import pymysql.cursors

class MySQLDB:
    """MySQL DB API"""

    def __init__(self,
                 host: str,
                 port: int,
                 user: str,
                 password: str,
                 database: str,
                 charset: str = "utf8mb4") -> None:
        """
        连接 MySQLDB
        :param host: 地址
        :param port: 端口
        :param user: 用户名
        :param password: 密码
        :param database: 数据库名称
        :param charset: 编码类型
```

```python
        """
        self.connection = pymysql.connect(
            host=host,
            port=int(port),
            user=user,
            password=password,
            database=database,
            charset=charset,
            cursorclass=pymysql.cursors.DictCursor)

    def execute_sql(self, sql: str) -> None:
        """
        执行 SQL 语句
        """
        with self.connection.cursor() as cursor:
            self.connection.ping(reconnect=True)
            if "delete" in sql.lower()[0:6]:
                cursor.execute("SET FOREIGN_KEY_CHECKS=0;")
            cursor.execute(sql)
        self.connection.commit()

    def query_sql(self, sql: str) -> list:
        """
        查询 SQL 语句
        """
        data_list = []
        with self.connection.cursor() as cursor:
            self.connection.ping(reconnect=True)
            cursor.execute(sql)
            rows = cursor.fetchall()
            for row in rows:
                data_list.append(row)
            self.connection.commit()
            return data_list

    def query_one(self, sql: str) -> Any:
        """
        查询 SQL 语句，返回一条数据
        """
        with self.connection.cursor() as cursor:
            self.connection.ping(reconnect=True)
            cursor.execute(sql)
            row = cursor.fetchone()
            self.connection.commit()
            return row
```

📖 代码说明

创建 MySQLDB 类，在 __init__() 初始化方法中设置连接数据库的配置。

- execute_sql()：执行 SQL 语句。
- query_sql()：查询 SQL 语句，其中，fetchall() 方法返回的查询结果为列表。如果查询结果为空，则返回空列表。
- query_one()：查询 SQL 语句，并通过 fetchone() 方法返回一条数据。如果有多条数据满足条件，那么只返回满足条件的第一条数据；如果没有数据满足条件，则返回 None。

使用示例

编写测试用例，验证 MySQLDB 类实现的方法是否可用。

```python
import unittest
from db_operate.mysql_db import MySQLDB

class MySQLTest(unittest.TestCase):
    """测试 MySQL 数据库操作 API"""

    def setUp(self):
        """初始化 DB 连接"""
        self.db = MySQLDB(
            host="localhost",
            port=3306,
            user="root",
            password="123456",
            database="guest")
        self.db.execute_sql("INSERT INTO api_user (name, age) VALUES ('test', 11) ")

    def test_execute_sql(self):
        """测试执行 SQL"""
        db = self.db
        db.execute_sql("INSERT INTO api_user (name, age) VALUES ('tom', 22) ")
        db.execute_sql("UPDATE api_user SET age=23 WHERE name='tom'")
        db.execute_sql("DELETE FROM api_user WHERE name = 'tom' ")
        result = db.query_sql("select * from api_user WHERE name='tom'")
        self.assertEqual(len(result), 0)

    def test_query_sql(self):
        """测试查询 SQL"""
        result = self.db.query_sql("select * from api_user")
        self.assertIsInstance(result, list)

    def test_query_one(self):
        """测试查询 SQL 一条数据"""
        result1 = self.db.query_one("select * from api_user")
        result2 = self.db.query_one("select * from api_user where id=99999")
        self.assertIsInstance(result1, dict)
        self.assertIsNone(result2)
```

执行结果如下。

```
> python test_mysql_api.py
...
----------------------------------------------------------------------
Ran 3 tests in 0.029s

OK
```

4.2.2 封装增查改删

在 4.2.1 节中,我们简化了 MySQL 的连接和操作,但仍然需要编写 SQL 语句。当字段非常多时,我们很难将字段名和字段值一一对应。比如,下面的这条插入 SQL 语句。

```
INSERT INTO
  table_name (
    `id`,
    `creator`,
    `reviser`,
    `createTime`,
    `shippingFee`,
    `totalAmount`,
    `sumProductPayment`,
    `currency`,
    `toFullName`,
    `toAddress`,
    `toFullAddress`,
    `storageName`,
    `orderTime`,
    `isSplit`,
    `packageNum`,
    `stockOutCreateTime`,
    `stockOutToFullName`,
    `stockOutToFullAddress`,
    `creatorName`,
    `stockOutId`,
    `orderId`
  ) VALUES (
    146,
    9002257,
    9002257,
    "2021-12-05 17:16:55",
    0,
    629,
    629,
    "RMB",
    "张三",
    "湖北省武汉市",
    "湖北省武汉市洪山区街道口",
    "初始仓库",
    "2021-12-05 17:16:55",
```

```
    0,
    "1/1",
    "2021-12-05 17:16:56",
    "张三",
    "湖北省武汉市洪山区街道口",
    "账号联系人",
    "1467422726779043840",
    "1467422722362441728"
)
```

如果将数据转换为 dict 格式,那么定义起来会更加简单,而且不容易出错,数据结构如下。

```
{
    "id": 146,
    "creator": 9002257,
    "reviser": 9002257,
    "createTime": "2021-12-05 17:16:55",
    "shippingFee": 0,
    "totalAmount": 629,
    "sumProductPayment": 629,
    "currency": "RMB",
    "toFullName": "张三",
    "toAddress": "湖北省武汉市",
    "toFullAddress": "湖北省武汉市洪山区街道口",
    "storageName": "初始仓库",
    "orderTime": "2021-12-05 17:16:55",
    "isSplit": 0,
    "packageNum": "1/1",
    "stockOutCreateTime": "2021-12-05 17:16:56",
    "stockOutToFullName": "张三",
    "stockOutToFullAddress": "湖北省武汉市洪山区街道口",
    "creatorName": "账号联系人",
    "stockOutId": "1467422726779043840",
    "orderId": "1467422722362441728"
}
```

基于这样的需求,我们可以进一步设计一套方法来实现数据的增、查、改、删。

实现 MySQL 语句的增、查、改、删方法。

```python
from typing import Any
import pymysql.cursors

class SQLBase:
    """SQL base API"""

    @staticmethod
    def dict_to_str(data: dict) -> str:
        """
        将字典转换为字符串,用逗号分隔
```

```python
        """
        tmp_list = []
        for key, value in data.items():
            if value is None:
                tmp = f"{key}=null"
            elif isinstance(value, int):
                tmp = f"{key}={value}"
            else:
                tmp = f"{key}='{value}'"
            tmp_list.append(tmp)
        return ','.join(tmp_list)

    @staticmethod
    def dict_to_str_and(conditions: dict) -> str:
        """
        将字典转换为字符串,用 and 连接
        """
        tmp_list = []
        for key, value in conditions.items():
            if value is None:
                tmp = f"{key}=null"
            elif isinstance(value, int):
                tmp = f"{key}={value}"
            else:
                tmp = f"{key}='{value}'"
            tmp_list.append(tmp)
        return ' and '.join(tmp_list)


class MySQLDB(SQLBase):
    """MySQL DB API"""

    def __init__(self,
                 host: str,
                 port: int,
                 user: str,
                 password: str,
                 database: str,
                 charset: str = "utf8mb4") -> None:
        ...

    def execute_sql(self, sql: str) -> None:
        ...

    def query_sql(self, sql: str) -> list:
        ...

    def query_one(self, sql: str) -> Any:
        ...

    def insert(self, table: str, data: dict) -> None:
```

```python
    """
    插入数据
    :param table: 表名
    :param data: 数据
    """
    for key in data:
        data[key] = "'" + str(data[key]) + "'"
    key = ','.join(data.keys())
    value = ','.join(data.values())
    sql = f"""insert into {table} ({key}) values ({value})"""
    self.execute_sql(sql)

def select(self, table: str, where: dict = None, one: bool = False) -> Any:
    """
    查询数据
    :param table: 表名
    :param where: 查询条件
    :param one: 是否返回一条数据
    """
    sql = f"""select * from {table} """
    if where is not None:
        sql += f""" where {self.dict_to_str_and(where)}"""
    if one is True:
        return self.query_one(sql)

    return self.query_sql(sql)

def update(self, table: str, data: dict, where: dict) -> None:
    """
    更新数据
    :param table: 表名
    :param data: 更新字段
    :param where: 查询条件
    """
    sql = f"""update {table} set """
    sql += self.dict_to_str(data)
    if where:
        sql += f""" where {self.dict_to_str_and(where)};"""
    self.execute_sql(sql)

def delete(self, table: str, where: dict = None) -> None:
    """
    删除数据
    :param table: 表名
    :param where: 查询条件
    """
    sql = f"""delete from {table}"""
    if where is not None:
```

```
            sql += f""" where {self.dict_to_str_and(where)};"""
        self.execute_sql(sql)
```

📖 代码说明

注：前面已经实现的方法，在这里用 "..." 表示。

首先，在 SQLBase 类中实现如下方法。

- dict_to_str()：将字典转换为字符串，并用逗号","分隔。比如，将{"id": 1, "name": "tom"}转换为"id=1, name=tom"。在更新 SQL 语句时需要用到这样的字符串。
- dict_to_str_and()：将字典转换为字符串，并用 and 连接。比如，将{"id": 1, "name": "tom"}转换为"id=1 and name=tom"。在 SQL 语句的 where 条件中需要用到这样的字符串。

其次，修改 MySQLDB 类，使其继承 SQLBase 类，并新增 SQL 操作方法。

- insert()：插入 SQL 语句。用 table 指定表名，用 data 指定插入的数据。
- select()：查询 SQL 语句。用 table 指定表名，用 where 指定查询的条件，用 one 指定是否返回单条数据。
- update()：更新 SQL 语句。用 table 指定表名，用 data 指定更新的数据，用 where 指定查询的条件。
- delete()：删除 SQL 语句。用 table 指定表名，用 where 指定查询的条件。

🔖 使用示例

编写测试用例，验证 MySQLDB 类实现的查、删、改、增方法是否可用。

```python
import unittest
from db_operate.mysql_db import MySQLDB

class MySQLTest(unittest.TestCase):
    """测试 MySQL 数据库操作 API"""

    def setUp(self):
        """初始化 DB 连接"""
        self.db = MySQLDB(
            host="localhost",
            port=3306,
            user="root",
            password="123456",
            database="guest")
        self.db.execute_sql("INSERT INTO api_user (name, age) VALUES ('test', 11) ")

    def tearDown(self):
        self.db.delete("api_user", {"name": "test"})

    def test_select_sql(self):
```

```python
        """测试查询 SQL"""
        result1 = self.db.select(table="api_user", where={"name": "test"})
        self.assertEqual(result1[0]["name"], "test")
        result2 = self.db.select(table="api_user", one=True)
        self.assertIsInstance(result2, dict)

    def test_delete_sql(self):
        """测试删除 SQL"""
        # delete sql
        self.db.delete(table="api_user", where={"name": "test"})
        result = self.db.query_sql("select * from api_user WHERE name='test'")
        self.assertEqual(len(result), 0)

    def test_update_sql(self):
        """测试更新 SQL"""
        self.db.update(table="api_user", where={"name": "test"}, data={"age": "22"})
        result = self.db.query_sql("select * from api_user WHERE name='test'")
        self.assertEqual(result[0]["age"], 22)

    def test_insert_sql(self):
        """测试插入 SQL"""
        data = {"name": "jean", "age": 11}
        self.db.insert(table="api_user", data=data)
        result = self.db.query_sql("select * from api_user WHERE name='jean'")
        self.assertTrue(len(result[0]) > 1)

if __name__ == '__main__':
    unittest.main()
```

使用封装的 select()、delete()、update()和 insert() 等方法会让数据库操作变得更加简单且有趣。但是这些方法并不支持复杂的 SQL 语句，比如联表查询，所以，设计它的目的并不是替代 SQL 语句，而是简化 SQL 操作。如果遇到复杂的 SQL 语句，那么请直接使用 execute_sql()方法和 query_sql() 方法。

如果在你的项目中需要用到其他数据库，那么同样可以根据 MySQLDB 类的设计思路封装数据库操作 API。

第 5 章 随机测试数据设计

测试数据的准备是自动化测试用例的重要部分，除与业务强相关的测试数据外，非强相关的测试数据可以随机生成。比如，注册新用户，注册时都要使用一个未注册过的账号；预订酒店日期查询，需要使用当天或未来的日期进行查询。类似的场景还有很多，需要通过生成随机数据来进行测试。

生成随机数据的方式有两种：

- 使用开源库：比如，Faker 就是非常强大的用来生成随机数据的工具，常见的数据都可以用它来生成。
- 自己封装：因为不同公司的业务不同，所以用到的测试数据也不同。为了满足自身业务需求，应实现对随机数据的封装。

5.1 测试工具介绍

5.1.1 Faker

Faker 是一个生成假数据的 Python 包，可用于快速生成各种类型的虚假数据，如姓名、地址和文本等。

用 pip 命令安装 Faker。

```
> pip install faker
```

使用示例

下面创建 faker_demo.py 文件，以了解 Faker 可以生成哪些类型的随机数据。

```
# faker_demo.py
from faker import Faker
```

```python
fake = Faker(locale='zh_CN')

# 基本常用数据
name = fake.name()
address = fake.address()
phone = fake.phone_number()
ID = fake.ssn(min_age=18)

print(f"""
    name: {name},
    address: {address},
    phone: {phone},
    ID: {ID}""")

# 与日期相关的数据
day_of_month = fake.day_of_month()
day_of_week = fake.day_of_week()
date = fake.date(pattern="%Y-%m-%d")
date_between = fake.date_between(start_date="-30y", end_date="today")
future_datetime = fake.future_datetime(end_date="+30d")
past_datetime = fake.past_datetime(start_date="-30d")

print(f"""
    day_of_month: {day_of_month},
    day_of_week: {day_of_week},
    data: {date},
    date_between: {date_between},
    future_datetime: {future_datetime},
    past_datetime: {past_datetime}""")

# 与网络相关的数据
email = fake.ascii_free_email()
domain = fake.domain_name(levels=1)
host = fake.hostname()
image_url = fake.image_url()
url = fake.url()
uri = fake.uri()
ipv4 = fake.ipv4(network=False)
ipv6 = fake.ipv6(network=False)

print(f"""
    email: {email},
    domain: {domain},
    host: {host},
    image_url: {image_url},
    url: {url},
    uri: {uri},
    ipv4: {ipv4},
    ipv6: {ipv6}""")
```

📖 代码说明

Faker 提供了非常多的数据类型，毫不夸张地说，它几乎可以生成任何你所需要的数据。这里只列举了一些比较常用的数据。

⏳ 运行结果

```
> python faker_demo.py

    name: 朱亮,
    address: 广东省广州市魏都永安街w座 876437,
    phone: 15184191514,
    ID: 513227195510311113

    day_of_month: 21,
    day_of_week: 星期四,
    data: 1986-05-01,
    date_between: 2022-05-15,
    future_datetime: 2023-07-13 02:25:00,
    past_datetime: 2023-05-31 18:01:50

    email: bgong@gmail.com,
    domain: 63.org,
    host: db-13.40.org,
    image_url: https://dummyi***.com/621x580,
    url: http://www.***.com/,
    uri: https://www.guiy***.cn/home.html,
    ipv4: 151.116.139.114,
    ipv6: e6fb:ae61:e172:8bfa:a2a3:8564:81a:eb6c
```

5.1.2 Hypothesis 库

Hypothesis 是一个支持多语言的测试库，它通过源示例来编写参数化测试。在执行时，它可以生成简单易懂的测试示例使测试失败，大大简化了测试用例的编写。通过它我们可以自动生成批量测试数据，摆脱手动构造数据这一烦琐任务，进而将精力集中在更高层次的测试逻辑上。

这种测试通常被称为"property-based testing"（基于属性的测试）。该概念最广为人知的实现是 Haskell 语言的 QuickCheck 库，但 Hypothesis 库与 QuickCheck 库有着明显的不同，Hypothesis 库的设计宗旨是适应你当前习惯的测试风格，你完全不需要熟悉 Haskell 语言或函数式编程语言。

Hypothesis 库目前支持 Python、Java 和 Ruby 等编程语言。

用 pip 命令安装 Hypothesis 库。

```
> pip install hypothesis
```

🖱 使用示例

使用 Hypothesis 库生成测试数据。

```python
# hypothesis_demo.py
import unittest
from hypothesis import given, settings
import hypothesis.strategies as st

def add(a: int, b: int):
    """被测试函数"""
    return a + b

class AddTest(unittest.TestCase):

    @settings(max_examples=10)
    @given(a=st.integers(), b=st.integers())
    def test_case(self, a, b):
        print(f"测试数据: a-> {a}, b-> {b}")
        c2 = add(a, b)
        self.assertIsInstance(c2, int)

if __name__ == '__main__':
    unittest.main()
```

📖 代码说明

settings 装饰器中的 max_examples 参数可以设置最多生成几组测试数据。

given 装饰器中的 a 和 b 参数通过使用 strategies 模块的 integers() 方法可以生成随机整数。

⏳ 运行结果

```
> python hypothesis_demo.py
测试数据: a-> 0, b-> 0
测试数据: a-> -79305309620772356774506035331520066215, b-> -49
测试数据: a-> 0, b-> 0
测试数据: a-> 0, b-> 0
测试数据: a-> -10470, b-> 1075412447
测试数据: a-> 10456, b-> 0
测试数据: a-> 10456, b-> 22679
测试数据: a-> 0, b-> 0
测试数据: a-> 0, b-> 0
测试数据: a-> 87, b-> -17406
.
----------------------------------------------------------------------
```

```
Ran 1 test in 0.010s

OK
```

通过测试结果，我们可以看到上面的代码随机生成了 10 组测试数据。

在使用 Hypothesis 库时，我们需要理解它的功能。

首先，虽然生成了 10 组测试数据，但是它们仍被记录为一个测试用例，因此，这 10 组测试数据中的任何一组测试数据都可能导致测试用例失败。

其次，因为测试数据是随机生成的，所以重复执行测试用例不一定能复现上次失败的问题。

最后，由于随机数据无法预知测试函数返回的结果，因此我们无法编写固定的断言数据，只能从类型或范围上进行断言，比如，断言整型、字符串类型等。

5.2 随机测试数据实战

除了使用现有的测试数据库，我们也可以根据自己的业务需求设计生成随机数据，这样的数据既更加轻量，也更加符合自己的业务测试需求。

📁 目录结构

```
05_chapter
├──testdata
│   └──testdata_func.py
└──test_testdata.py
```

5.2.1 随机生成手机号

手机号是很常用的一种数据，我国的手机号分为移动、联通和电信三大运营商，接下来实现随机生成手机号的方法。

```python
import re
import random

# 移动号段
mobile = [134, 135, 136, 137, 138, 139, 147, 150, 151, 152, 157, 158, 159, 172, 178,
          182, 183, 184, 187, 188, 195, 197, 198]
# 联通号段
unicom = [130, 131, 132, 145, 155, 156, 166, 175, 176, 185, 186, 196]
# 电信号段
telecom = [133, 149, 153, 180, 181, 189, 173, 177, 190, 191, 193, 199]


def get_phone(operator: str = None) -> str:
    """
```

```
    get phone number
    :param operator: 指定运营商
    """
    if operator is None:
        all_operator = mobile + unicom + telecom
        top_third = random.choice(all_operator)
    elif operator == "mobile":
        top_third = random.choice(mobile)
    elif operator == "unicom":
        top_third = random.choice(unicom)
    elif operator == "telecom":
        top_third = random.choice(telecom)
    else:
        raise TypeError("Please select the right operator: 'mobile', 'unicom', 'telecom' ")

    suffix = random.randint(9999999, 100000000)

    return f"{top_third}{suffix}"
```

代码说明

我国的手机号由前 3 位号段和后 8 位数字组成。首先，找出移动、联通、电信各自的号段。

根据 operator 参数，random.choice()方法可以随机从一个列表中取出元素。如果不选择运营商，那么将随机从三家运营商的号段中取出一个号段。

后 8 位数字非常简单，先使用 random.randint()方法设置数字范围，再随机生成一个整数即可。

使用示例

调用前文实现的随机生成手机号的方法，验证功能。

```
# test_testdata.py
from testdata.testdata_func import get_phone

# 随机生成手机号
print("手机号:", get_phone())
print("手机号(移动):", get_phone(operator="mobile"))
print("手机号(联通):", get_phone(operator="unicom"))
print("手机号(电信):", get_phone(operator="telecom"))
```

运行结果

```
> python test_testdata.py
手机号: 15819717787
手机号(移动): 19873466525
```

手机号(联通)：16651698530
手机号(电信)：17733753797

5.2.2　随机生成中文姓名

中文姓名是比较常用的一种数据，接下来实现随机生成中文姓名的方法。

```
import re
import random

zh_names_male = list(set(re.split(r"\s+", """德义 苍 鹏云 炎 和志 新霁 澜 星泽 驰轩 楚 宏
深 全 波涛 飞文 波 振国 凯 光启 经略 乐天 志强 作人 英叡 英华 星阑 景龙 鹏鲸 采 浩然 举 芬 鸿才
卫 嘉纳 旭东 玉泽 祺瑞 荫 茂德 博 鸿羲 彦 涵衍 开诚 鸿远 凯歌 星华 玉宇 潍 德华 甲 梓 正阳 文乐
高杰 骁 腾逸 鸿畅 修平""".strip()))

zh_names_female = list(set(re.split(r"\s+", """海莹 曼珠 虹影 凝安 淳美 清润 旋 馨香 骊
霞 水丹 长文 怀薇 平卉 向露 秀敏 青柏 尔阳 奥婷 智美 雅可 骊燕 燕珺 白曼 春枫 谷之 暖姝 易绿 娅欣
欢 半梅 忆彤 宇 茗 芳洁 双文 艳芳 珍丽 杨 若星 松葳 晓畅 菱华 新荣 觅露 冰夏 初柳 迎蕾 海宁 香
妙颜 靖之""".strip()))

zh_last_name = list(set(re.split(r"\s+", """赵 钱 孙 李 周 吴 郑 王 冯 陈 褚 卫 蒋 沈 韩
杨 朱 秦 尤 许 何 吕 施 张 孔 曹 严 华 宇文 尉迟 延陵 羊舌 欧阳 长孙 上官 司徒 司马 夏侯 西门 南
宫 公孙""".strip())))

def first_name(gender: str = "") -> str:
    """
    generate first name
    :param gender:
    :return:
    """
    genders = ["", "m", "f", "male", "female"]
    if gender not in genders:
        raise ValueError("Unsupported gender, try [m, f, male, female] instead")

    if gender == "":
        return random.choice(zh_names_female + zh_names_male)
    elif gender == "m":
        return random.choice(zh_names_male)
    else:
        return random.choice(zh_names_female)


def last_name() -> str:
    """
    generate last name
    :return:
    """
    return random.choice(zh_last_name)
```

```python
def name() -> str:
    """
    generate Chinese name
    """
    return f"{last_name()}{first_name()}"
```

📖 **代码说明**

首先，准备一些中文姓名的数据，可以到中文取名网站搜索。这里只是为了演示功能，准备了少量的"姓"和"名字"的数据。

first_name()函数可随机生成"名字"，可以指定性别为男或女。

last_name()函数可随机生成"姓"。

name()函数可随机生成"姓名"。

🖱 **使用示例**

调用前面实现的随机生成中文姓名的方法，验证功能。

```python
from testdata.testdata_func import first_name, last_name, name

# 随机生成一个名字
print("名字: ", first_name())
print("名字(男): ", first_name(gender="male"))
print("名字(女): ", first_name(gender="female"))

# 随机生成一个姓
print("姓:", last_name())

# 随机生成一个姓名
print("姓名:", name())
```

⏱ **运行结果**

```
> python test_testdata.py
名字：馨香
名字(男)：旋
名字(女)：春枫
姓：周
姓名：蒋初柳
```

5.2.3 获取在线时间

在实际测试过程中经常会遇到时间的问题，比如获取当前时间来计算商品的过期时间。由于程序在测试时获取的是本地系统的时间，即北京时间，所以在测试环境中测试用例可以顺利通过。但是当把测试用例部署到服务器上运行时，由于获取的是服务器时间，而服务器时间比北京时间早 8 小时，因此导致测试用例运行失败。

基于以上需求，接下来实现获取在线时间的方法。

```python
import time
import datetime
import requests

TAOBAO_TIME = "http://api.m.tao***.com/rest/api3.do?api=mtop.common.getTimestamp"

def online_timestamp() -> str:
    """
    get now timestamp
    :return:
    """
    r = requests.get(TAOBAO_TIME)
    data = r.json()
    ts = data["data"]["t"]
    return ts


def online_datetime() -> [str, datetime]:
    """
    get online date time
    :return:
    """
    ts = online_timestamp()
    date_time = time.strftime("%Y-%m-%d %H:%M:%S", time.localtime(int(ts[:10])))
    return date_time
```

📖 **代码说明**

首先，使用淘宝开放的一个 API 来获取时间，类似免费的 API 还有很多。其次，通过 requests 库调用该 API 来获取时间。第 10 章会重点介绍 requests 库的使用和封装。

- online_timestamp()方法用于获取当前时间戳。
- online_datetime()方法用于将时间戳转为日期时间格式。

💡 **使用示例**

调用前面实现的获取在线时间的方法，验证功能。

```python
from testdata.testdata_func import online_timestamp, online_datetime

# 获取在线时间
print("当前时间戳", online_timestamp())
```

```
print("当前日期时间", online_datetime())
```

运行结果

```
> python test_testdata.py
当前时间戳 1709656271193
当前日期时间 2024-03-06 00:31:11
```

以上这些随机数据生成方法仅仅起到抛砖引玉的作用，在日常自动化测试过程中，你可能会遇到各种需求的测试数据，这时你就可以结合上面的方法进行封装，以随机生成适应特定业务场景的数据。

第 6 章 命令行工具设计

命令行工具可以有效地提高生产效率，在日常工作中命令几乎无处不在。比如，操作系统下的 shell 命令、编程语言的运行命令，以及各种开发测试工具和框架的执行命令。当然，我们也可以设计和实现命令行工具。

6.1 用 Python 实现命令行工具

在 Python 中实现命令行工具的方式有多种。
- argv：Python 提供的一种简单的标准库，通过 sys 的 argv 可以接收命令行参数。
- argparse：Python 内置的用于解析命令行选项与参数的模块。
- 第三方库：Python 有很多简单好用的 CLI 库，比如，click、python-fire 和 typer。

6.1.1 argv 的使用

sys 是 Python 的一个标准库，它是 System 的简写，封装了一些系统的信息和接口。argv 是 argument variable 的简写，一般在命令行调用时由系统传递给程序。

🖱 使用示例

```
# argv_demo.py
import sys

run_file = sys.argv[0]
print(f"file name -> {run_file}")

params = sys.argv[1:]
for i in params:
```

```python
    print(f"hello, {i}")
```

📖 代码说明

argv 其实是一个 List 列表，argv[0]一般是被调用的脚本文件名，argv[1:] 表示在文件名之后传入的参数。

⏳ 运行结果

```
> python argv_demo.py tom jack
file name -> argv_demo.py
hello, tom
hello, jack
```

6.1.2 argparse 的使用

argparse 是 Python 内置的用于解析命令行选项和参数的模块，argparse 可以让我们轻松地编写用户友好的命令行接口。

使用示例

```python
# argp_demo.py
import argparse

parser = argparse.ArgumentParser(description='argparse 简单用法')

parser.add_argument('-n', '--name', type=str, default="tom", help="请输入名字，默认为 tom")
parser.add_argument('-c', '--count', type=int, default=1, help="请输入次数，默认值为 1")

args = parser.parse_args()

# 使用参数
name = args.name
count = args.count
for _ in range(count):
    print(f"hello, {name}")
```

📖 代码说明

ArgumentParser 类定义了命令行解析对象。add_argument()方法添加了命令行参数。

- -n，--name：指定参数名。
- type：指定参数类型。
- default：设置默认值。

- help：定义参数说明信息。

parse_args()方法首先从命令行中结构化解析参数，然后通过 args.××× 得到具体的参数，最后基于参数完成相关功能的开发。

📖 查看帮助

```
> python argp_demo.py --help
Usage: argp.py [-h] [-n NAME] [-c COUNT]

argparse 的简单用法

Options:
  -h, --help            show this help message and exit
  -n NAME, --name NAME  请输入名字，默认为 tom
  -c COUNT, --count COUNT  请输入次数，默认值为 1
```

⏳ 运行结果

```
> python argp_demo.py --count 3 --name jack
hello, jack
hello, jack
hello, jack
```

6.1.3　click 的用法

click 是一个 Python 工具包，用于以可组合的方式创建美观的命令行界面，并允许开发者以尽可能少的代码实现命令功能。此外，它是高度可配置的，并附带了合理的默认值。

用 pip 命令安装 click。

```
> pip install click
```

📖 使用示例

```python
# click_demo.py
import click

@click.command()
@click.option("-c", "--count", default=1, help="执行次数，默认值为1。")
@click.option("-n", "--name", prompt="Your name", help="请输入名字。")
def hello(count, name):
    """简单的CLI，问候 count 次 name。"""
    for _ in range(count):
        click.echo(f"Hello, {name}!")

if __name__ == '__main__':
    hello()
```

📖 代码说明

click 实现的功能与前面 argparse 实现的功能相同，但是 click 的用法显然更加简捷。

click.command 装饰器：将函数转换为命令行接口。

click.option 装饰器：添加命令行选项。

- -c, --cont：指定参数名。
- default：设置默认值。
- prompt：当参数为空时，提示输入，类似 input() 的用法。
- help：定义帮助信息。

🔎 查看帮助

```
> python click_demo.py --help
Usage: click_demo.py [OPTIONS]

  简单的 CLI，问候 count 次 name.

Options:
  -c, --count INTEGER  执行次数，默认值为 1
  -n, --name TEXT      请输入名字
  --help               显示此信息并退出
```

⏳ 运行结果

```
> python click_demo.py -c 3 -n tom
Hello, tom!
Hello, tom!
Hello, tom!

> python click_demo.py -c 3
Your name: jack   <-- 需要手动输入名字
Hello, jack!
Hello, jack!
Hello, jack!
```

6.2 命令行工具实战

通过前面的学习，我们已对命令行工具建立了基本的认识与理解。命令行工具的应用非常广泛，本节我们基于 click 制作两个命令行工具。

📁 目录结构

```
06_chapter
```

```
├──tools
│   ├──scaffolding.py
│   └──kb.py
```

6.2.1 实现自动化测试项目脚手架

做过 Web 自动化测试项目的同学应该知道,在创建自动化测试项目时,会有比较固定的目录结构,比如,自动化测试项目的目录结构。

📂 **目录结构**

```
mypro/
├──reports       # 测试报告目录
├──test_data     # 测试数据目录
├──test_case     # 测试用例目录
│   └──test_sample.py
└──run.py        # 运行入口文件
```

我们可以用脚手架来自动生成自动化测试项目。接下来,我们用 click 开发一个自动化测试项目的脚手架。

```python
# tools/scaffolding.py
import os
import click


@click.command()
@click.option("-P", "--project", help="指定自动化测试项目的项目名称")
def main(project):
    """自动化测试项目的脚手架"""
    if project:
        create_scaffold(project)
        return 0


def create_scaffold(project_name: str) -> None:
    """
    创建脚手架
    :param project_name: 项目名称
    :return:
    """

    if os.path.isdir(project_name):
        print(f"Folder {project_name} exists, please specify a new folder name.")
        return

    print(f"Start to create new test project: {project_name}")
    print(f"CWD: {os.getcwd()}\n")

    def create_folder(path: str):
```

```python
    """
    创建目录
    :param path: 路径
    :return:
    """
    os.makedirs(path)
    print(f"created folder: {path}")

def create_file(path: str, file_content: str = ""):
    """
    创建文件
    :param path: 路径
    :param file_content: 文件内容
    :return:
    """
    with open(path, 'w', encoding="utf-8") as py_file:
        py_file.write(file_content)
    msg = f"created file: {path}"
    print(msg)
# 测试用例模板
test_sample = '''import unittest


class MyTest(unittest.TestCase):

    def test_case(self):
        self.assertEqual(2+2, 4)


if __name__ == '__main__':
    unittest.main()

'''
    # 运行测试脚本
    run = '''import unittest

suit = unittest.defaultTestLoader.discover("test_dir", "test_*.py")

runner = unittest.TextTestRunner()
runner.run(suit)

'''
    create_folder(project_name)
    create_folder(os.path.join(project_name, "test_dir"))
    create_folder(os.path.join(project_name, "reports"))
    create_folder(os.path.join(project_name, "test_data"))
    create_file(os.path.join(project_name, "test_dir", "__init__.py"))
    create_file(os.path.join(project_name, "test_dir", "test_sample.py"),
test_sample)
```

```
        create_file(os.path.join(project_name, "run.py"), run)


if __name__ == '__main__':
    main()
```

📖 **代码说明**

-P/--project 参数：指定自动化测试项目的项目名称。

create_scaffold()函数：创建脚手架。其中，create_folder()子函数用于创建目录，create_file()子函数用于创建文件。

文件中的测试代码是基于 unittest 实现的，比较简单。

🔖 **查看帮助**

```
> python scaffolding.py --help
Usage: scaffolding.py [OPTIONS]

  自动化测试项目的脚手架

Options:
  -P, --project TEXT  指定自动化测试项目的项目名称
  --help              显示此条信息并退出
```

⏳ **运行结果**

```
> python scaffolding.py -P mypro
Start to create new test project: mypro
CWD: D:\github\test-framework-design\06_chapter\tools

created folder: mypro
created folder: mypro\test_dir
created folder: mypro\reports
created folder: mypro\test_data
created file: mypro\test_dir\__init__.py
created file: mypro\test_dir\test_sample.py
created file: mypro\run.py
```

运行完成，我们会在当前目录中得到一个名为"mypro"的项目，目录结构见本小节开头部分，它包含了自动化测试项目用到的目录和文件。

6.2.2 实现性能工具

许多性能测试工具都提供了命令行模式，或者本身就以命令行形式存在。比如，Apache ab（Apache HTTP server benchmarking tool）和 locust。

下面我们通过 Python 设计一个命令行性能测试工具。这个工具稍微有些复杂，涉及以下几个库。

- gevent：gevent 是一个基于协程的 Python 网络库，它使用 greenlet 在 libev 或 libuv 事件循环之上提供高级同步 API。性能工具的核心就是模拟用户并发请求，当通过 gevent 模拟用户并发请求时，在同等硬件配置下可以有更好的性能。
- requests：一个 HTTP(s)客户端库，用于发送 HTTP 请求。当通过性能测试模拟用户并发请求时，其本质是向服务器端发送请求。
- tqdm：用于显示进度条的一个库。在性能测试运行过程中，通过 tqdm 可以直观地看到运行进度。
- numpy：Python 中一种开源的数值计算扩展库，用它来统计测试结果非常方便。

用 pip 命令安装依赖库。

```
> pip install gevent
> pip install requests
> pip install numpy
> pip install tqdm
```

完整的实现代码如下。

```python
# tools/kb.py
import gevent
from gevent import monkey
monkey.patch_all()
import time
import click
import requests
from numpy import mean
from tqdm import tqdm


class Statistical:
    """统计类"""
    pass_number = 0
    fail_number = 0
    run_time_list = []


def running(url, request):
    """运行请求调用"""
    for _ in tqdm(range(request)):
        start_time = time.time()
        r = requests.get(url)
        if r.status_code == 200:
            Statistical.pass_number = Statistical.pass_number + 1
        else:
            Statistical.fail_number = Statistical.fail_number + 1
```

```python
        end_time = time.time()
        run_time = round(end_time - start_time, 4)
        Statistical.run_time_list.append(run_time)


@click.command()
@click.argument('url')
@click.option('-u', '--user', default=1, help='运行用户的数量，默认值为1', type=int)
@click.option('-q', '--request', default=1, help='单个用户请求数，默认值为 1', type=int)
def main(url, user, request):
    """指定请求的URL"""
    print(f"请求URL: {url}")
    print(f"用户数: {user}, 循环次数: {request}")
    print("============== Running ==================")

    jobs = [gevent.spawn(running, url, request) for _url in range(user)]
    gevent.wait(jobs)

    print("\n============== Results ==================")
    print(f"最大:      {str(max(Statistical.run_time_list))} s")
    print(f"最小:      {str(min(Statistical.run_time_list))} s")
    print(f"平均:      {str(round(mean(Statistical.run_time_list), 4))} s")
    print(f"请求成功次数: {Statistical.pass_number}")
    print(f"请求失败次数: {Statistical.fail_number}")
    print("=================== end ====================")


if __name__ == "__main__":
    main()
```

📖 代码说明

首先，gevent 导入顺序必须放到最前面，monkey.patch_all() 用于解决协程堵塞问题。

当通过 click 实现命令行工具时，需要接收三个参数。

- url 参数：用于指定性能测试的 URL 地址，默认为第一个且必须上传。
- user 参数：用于设置运行用户的数量（即并发数）。
- request 参数：用于设置单个用户请求数。

gevent.spawn() 用于创建一个协程并运行，它可以根据 user 的数量创建相应数量的协程。

running() 函数使用 requests 库，可以根据 request 数量发送请求，并判断返回的 HTTP 状态码是否为 200。

最后，针对请求的次数、时间进行统计，分别计算最大、最小、平均请求时间，以及总的请求成功次数、请求失败次数等。

查看帮助如下。

```
> python kb.py --help
Usage: kb.py [OPTIONS] URL
    指定请求的 URL
Options:
  -u, --user INTEGER     运行用户的数量，默认值为 1
  -q, --request INTEGER  单个用户请求数，默认值为 1
  --help                 Show this message and exit.
```

📄 运行结果

```
> python kb.py https://www.ba***.com -u 2 -q 10
请求 URL: https://www.ba***.com
用户数: 2, 循环次数: 10
============== Running ==================
100%|████████████| 10/10 [00:02<00:00, 4.33it/s]
100%|████████████| 10/10 [00:02<00:00, 4.00it/s]
============== Results ==================
最大:       0.4874 s
最小:       0.078 s
平均:       0.2401 s
请求成功次数: 20
请求失败次数: 0
================= end ===================
```

6.2.3 生成命令行工具

到目前为止，我们虽然可以利用 click 实现类似的命令行工具，但是真正的命令行工具允许在终端任意位置使用"×××"命令，而不是"python ×××.py"命令。也就是说，可以脱离"python"的前缀来使用命令。

参考 1.6 节，使用 Poetry 创建一个项目，通过安装的方式生成命令行工具。

```
> poetry new kb
```

📁 目录结构

```
kb/
├──pyproject.toml
├──README.md
├──kb
│   ├──cli.py
│   └──__init__.py
```

```
└─tests
    └─__init__.py
```

手动添加 cli.py 文件，在文件中实现命令行工具。可以直接复制 tools/kb.py 文件中的代码到 kb/cli.py 文件中。

在 pyproject.toml 文件中设置命令行工具。

```
[tool.poetry]
name = "kb"
version = "0.1.0"
description = "kb is a simple performance testing tool"
authors = ["fnngj <fnngj@126.com>"]
readme = "README.md"

[tool.poetry.dependencies]
python = "^3.11"
requests = "^2.32.3"
gevent = "^24.2.1"
numpy = "^2.0.0"
tqdm = "^4.66.4"

[tool.poetry.scripts]
kb = "kb.cli:main"

[build-system]
requires = ["poetry-core"]
build-backend = "poetry.core.masonry.api"
```

为了使 pyproject.toml 文件更加简洁，这里省略了对一些参数的定义，比如，Home-page、License、Requires 和 Required-by。我们可以参考第 1 章中对 pyproject.toml 文件的定义进行设置。

[tool.poetry.scripts] 用于设置命令行工具，kb 为命令行工具名称。比如，"kb.cli.main" 中的 kb 为 "kb/" 目录，cli 为 "cli.py" 文件，main 为 "main()" 方法。

设置后，在 kb 项目的根目录下，使用 pyproject.toml 文件安装 kb 项目。

```
> pip install .
```

下面通过 pip 命令查看安装信息。

```
> pip show kb
Name: kb
Version: 0.1.0
Summary: kb is a simple performance testing tool
Home-page:
Author: defnngj
Author-email: fnngj@126.com
License:
Location: C:\Python311\Lib\site-packages
Requires:
Required-by:
```

使用示例

安装成功后,我们就可以在终端的任意位置使用 kb 命令了。

```
> kb --help
Usage: kb [OPTIONS] URL

Options:
  -u, --user INTEGER     运行用户的数量,默认值为 1
  -q, --request INTEGER  单个用户请求数,默认值为 1
  --help                 Show this message and exit.
```

至此,我们就设计出一个名为"kb"的命令行性能测试工具了,值得庆祝一下!

第 7 章 测试框架扩展功能设计

unittest 作为 Python 默认的单元测试框架,虽然用法比较简单,但是功能不够强大,扩展库也比较少,本章将介绍基于 unittest 的常用扩展功能的开发方法。

7.1 测试用例依赖

在设计自动化测试用例时,我们应尽量避免测试用例之间的依赖,即每条测试用例都可以独立执行。但是,有时使用依赖可以有效地简化测试用例的编写。

使用测试用例依赖一般分为两种情况。

- 依赖测试用例结果:根据被依赖测试用例的失败、错误或跳过等结果,决定是否执行当前测试用例。
- 依赖测试条件:根据一定的条件,决定是否依赖测试用例。

📂 目录结构

```
07_chapter/
├──extends
│   ├──depend_extend.py
│   └──__init__.py
├──test_depend.py
└──test_if_depend.py
```

7.1.1 依赖测试用例结果

我们可以设计一个装饰器,实现测试用例依赖。

```python
# extends/depend_extend.py
import functools
```

```python
from unittest import skipIf

def depend(case=None):
    """
    depend装饰器
        :param case: 依赖的测试用例名称
        :return:
    """
    def wrapper_func(test_func):
        @functools.wraps(test_func)
        def inner_func(self, *args):
            if case == test_func.__name__:
                raise ValueError(f"{case} cannot depend on itself")
            failures = str([fail_[0] for fail_ in self._outcome.result.failures])
            errors = str([error_[0] for error_ in self._outcome.result.errors])
            skipped = str([skip_[0] for skip_ in self._outcome.result.skipped])
            flag = (case in failures) or (case in errors) or (case in skipped)
            test = skipIf(flag, f'{case} failed or error or skipped')(test_func)
            try:
                return test(self)
            except TypeError:
                return None
        return inner_func
    return wrapper_func
```

📖 **代码说明**

实现 depend 装饰器，test_func.__name__ 可以获取被装饰方法的名称，之后进行判断。如果该名称等于依赖的测试用例名称，则抛出异常。简单来说，就是不允许测试用例自己依赖自己。

self._outcome.result 可以获取在 TextTestResult 类中已经运行的测试结果，目的是找出失败（failures）、错误（errors）和跳过（skipped）的测试用例，并判断被依赖的测试用例结果是否属于以上三种。如果是，则将依赖的测试用例设置为跳过。

举一个简单的例子，在你的日常工作中有一项工作需要依赖同事小张，因此在你进行这项工作之前，你会判断小张的工作是否完成。如果小张的工作没有完成，那么你的这项工作也不用做。

🔖 **使用示例**

下面通过示例演示 depend 装饰器的具体用法。

```python
import unittest
from extends.depend_extend import depend

class TestDepend(unittest.TestCase):
```

```python
    def test_001(self):
        print("test_001")
        self.assertEqual(1+1, 3)

    @depend("test_001")
    def test_002(self):
        print("test_002")

    @depend("test_002")
    def test_003(self):
        print("test_003")


if __name__ == '__main__':
    unittest.main()
```

📖 **代码说明**

测试用例 test_003 依赖于测试用例 test_002，测试用例 test_002 又依赖于测试用例 test_001。当被依赖的测试用例为错误、失败或跳过时，该测试用例可自动跳过。

这里有一个前提，即被依赖的测试用例一定要先被执行。如何保证被依赖的测试用例先被执行？我们可以根据 unittest 查找测试用例的规则，通过给测试方法命名进行编号排序。比如，在同一个测试类下面，测试用例 test_a 优先于测试用例 test_b 执行，测试用例 test_1 优先于测试用例 test_2 执行。除此之外，我们可以使用 unittest.TestSuite()手动添加测试用例，使测试用例按照被添加的顺序执行，代码如下。

```python
...

if __name__ == '__main__':
    # 手动控制测试用例的添加顺序
    suite = unittest.TestSuite()
    suite.addTests([
        TestDepend("test_001"),
        TestDepend("test_002"),
        TestDepend("test_003"),
    ])
    runner = unittest.TextTestRunner()
    runner.run(suite)
```

⏳ **运行结果**

```
> python test_depend.py
test_001
Fss
======================================================================
FAIL: test_001 (__main__.TestDepend.test_001)
----------------------------------------------------------------------
Traceback (most recent call last):
```

```
  File "D:\github\test-framework-design\07_chapter\test_depend.py", line 9, in
test_001
    self.assertEqual(1+1, 3)
AssertionError: 2 != 3

----------------------------------------------------------------------
Ran 3 tests in 0.001s

FAILED (failures=1, skipped=2)
```

通过运行结果可以看出，测试用例 test_001 断言失败，导致测试用例 test_002 被跳过。而测试用例 test_003 依赖于测试用例 test_002，所以测试用例 test_003 也被跳过。整个运行结果是：1 条失败，2 条跳过。

7.1.2　依赖测试条件

有些时候，被依赖的测试用例可以手动控制依赖的测试用例是否执行。

同样以工作为例，在你的日常工作中有一项工作需要依赖同事小张，因此在你进行这项工作之前，你会根据小张反馈的结果来决定是否开启这项工作。注意，这里以小张的意见为准，而不是以小张工作本身的结果为准。

```python
# extends/depend_extend.py

def if_depend(value):
    """
    根据条件跳过当前测试用例
    :param value:变量的值
    :return:
    """
    def wrapper_func(function):
        def inner_func(self, *args, **kwargs):
            if not getattr(self, value):
                self.skipTest('Dependent use case not passed')
                pass
            else:
                function(self, *args, **kwargs)
        return inner_func
    return wrapper_func
```

📖 **代码说明**

if_depend 装饰器的实现相对比较简单，即通过判断 value 的结果决定是否跳过依赖的测试用例。如果 value 为 False，则调用 self.skipTest 跳过依赖的测试用例。如果 value 为 True，则依赖的测试用例正常执行。

> **使用示例**

下面通过示例演示 if_depend 装饰器的具体用法。

```python
import unittest
from extends.depend_extend import if_depend

class TestIfDepend(unittest.TestCase):
    depend_flag = True

    def test_001(self):
        TestIfDepend.depend_flag = False  # 修改 depend_flag 为 False

    @if_depend("depend_flag")
    def test_002(self):
        ...

if __name__ == '__main__':
    unittest.main()
```

在测试类中设置全局变量 depend_flag 的值为 True，在测试用例 test_001 中根据情况修改 depend_flag 的值。在测试用例 test_002 中通过 if_depend 装饰器取 depend_flag 的值进行判断。注意，depend_flag 是作为字符串传入@if_depend()装饰器的。

> **运行结果**

```
> python test_if_depend.py
.s
----------------------------------------------------------------------
Ran 2 tests in 0.000s

OK (skipped=1)
```

通过执行结果可以看到，在测试用例 test_001 中，修改 depend_flag 的值为 False 后，导致测试用例 test_002 被跳过。

7.2 测试用例分类标签

一般情况下，我们会按照功能模块的维度对测试用例进行划分。比如，一个测试平台通常包含登录（注册）、项目管理、用例管理和任务管理等模块，在编写自动化测试用例时，我们会按照这个维度来创建对应的目录和文件结构。在运行测试时，也是按照功能模块来进行的。

除此之外，同一模块中的测试用例也有其他维度的划分，比如，优先级、版本号。以优先级为例，在登录测试用例中，合法用户登录成功的测试用例优先级最高；在项目管理测试用例中，合法数据创建项目成功的测试用例优先级最高。这些都是验证功能是否可用的最基本的测试用例。

因此，我们在执行测试用例时，有可能为了节约执行时间而选择只运行优先级最高的测试用例。而测试用例的存放方式只能以一种维度保存：功能模块、优先级或其他。这就引申出一个问题，即如何按照不同的维度找出某一类测试用例，并执行它们？给测试用例打上不同的标签就是一种解决方案。

7.2.1 实现分类标签

以 unittest 为例，为了让 unittest 支持标签的功能，我们需要分两步来实现。

第一步，实现测试用例标签装饰器。

第二步，重写 unittest 的 TextTestRunner 类的部分方法，识别测试用例标签。

📂 目录结构

```
07_chapter/
├──extends
│   ├──label_extend.py
│   └──__init__.py
├──runner
│   └──runner.py
└──test_label.py
```

实现测试用例标签装饰器。

```python
# extends/label_extend.py
def label(*labels):
    """
    测试用例分类标签
    """

    def inner(cls):
        """
        为类或方法添加标签
        """
        cls._labels = set(labels) | getattr(cls, '_labels', set())
        return cls

    return inner
```

📖 代码说明

实现 label 装饰器，为类或方法添加分类标签，其中，set()函数用于创建一个无序且不重复的元素集，getattr() 函数用于返回一个对象属性值。

接下来，继承并重写 unittest 的 TextTestRunner 类的部分方法。

```python
# runner/runner.py
import unittest
import functools

class MyTestRunner(unittest.TextTestRunner):
    """
    继承并重写 TextTestRunner 类的部分方法
    """
    def __init__(self, *args, **kwargs):
        """
        增加属性:
        * 黑名单（blacklist）
        * 白名单（whitelist）
        """
        self.whitelist = set(kwargs.pop('whitelist', []))
        self.blacklist = set(kwargs.pop('blacklist', []))

        super(MyTestRunner, self).__init__(*args, **kwargs)

    @classmethod
    def test_iter(cls, suite):
        """
        遍历测试套件，生成单个测试用例
        """
        for test in suite:
            if isinstance(test, unittest.TestSuite):
                for t in cls.test_iter(test):
                    yield t
            else:
                yield test

    def run(self, testlist):
        """
        运行给定的测试用例或测试套件
        """
        suite = unittest.TestSuite()

        # 在测试列表中添加所有的测试用例，必要时使用跳过机制
        for test in self.test_iter(testlist):
            # 获取测试用例是否添加了标签，以及标签是否在白名单或黑名单中
            skip=bool(self.whitelist)
            test_method = getattr(test, test._testMethodName)
            test_labels = getattr(test, '_labels', set()) | getattr(test_method, '_labels', set())
            if test_labels & self.whitelist:
                skip = False
```

```
        if test_labels & self.blacklist:
            skip = True

    if skip:
        # 针对跳过的测试用例，用 skip 替换原始的方法
        @functools.wraps(test_method)
        def skip_wrapper(*args, **kwargs):
            raise unittest.SkipTest('label exclusion')
        skip_wrapper.__unittest_skip__ = True
        if len(self.whitelist) >= 1:
            skip_wrapper.__unittest_skip_why__ = f'label whitelist {self.whitelist}'
        if len(self.blacklist) >= 1:
            skip_wrapper.__unittest_skip_why__ = f'label blacklist {self.blacklist}'
        setattr(test, test._testMethodName, skip_wrapper)

    suite.addTest(test)

super(MyTestRunner, self).run(suite)
```

📖 代码说明

由于这部分代码比较复杂，所以笔者在代码中主要改动的位置添加了注释。

首先，在 __init__() 初始化方法中定义黑名单和白名单变量。

- 黑名单：需要跳过的测试用例。
- 白名单：需要运行的测试用例。

然后，新增 test_iter() 方法，遍历测试套件，并生成测试用例。

最后，重写 run() 方法，遍历 test_iter() 方法生成的测试用例，获取类或方法的标签。如果标签在白名单中，则不跳过测试用例（skip=False）；如果标签在黑名单中，则跳过测试用例（skip=True）。针对跳过的测试用例，用 skip 替换原始的方法。

7.2.2 使用分类标签

下面通过示例验证分类标签的功能是否可用。

🔖 使用示例

利用 label 装饰器和 MyTestRunner 类运行测试用例。

```
import unittest
from extends.label_extend import label
from runner.runner import MyTestRunner

class MyTest(unittest.TestCase):
```

```python
    @label("base")
    def test_label_base(self):
        self.assertEqual(1+1, 2)

    @label("slow")
    def test_label_slow(self):
        self.assertEqual(1, 2)

    def test_no_label(self):
        self.assertEqual(2+3, 5)

if __name__ == '__main__':
    suit = unittest.TestSuite()
    suit.addTests([
        MyTest("test_label_base"),
        MyTest("test_label_slow"),
        MyTest("test_no_label")
    ])
    runner = MyTestRunner(whitelist=["base"], verbosity=2)      # 白名单
    # runner = MyTestRunner(blacklist=["slow"], verbosity=2)    # 黑名单
    runner.run(suit)
```

📖 **代码说明**

我们分别编写三条测试用例：在第 1 条测试用例中设置标签为"base"，在第 2 条测试用例中设置标签为"slow"，在第 3 条测试用例中不设置标签。

⏳ **运行结果**

- 设置白名单为"base"。

```
> python test_label.py  # 白名单
test_label_base (__main__.MyTest.test_label_base) ... ok
test_label_slow (__main__.MyTest.test_label_slow) ... skipped "label whitelist
{'base'}"
test_no_label (__main__.MyTest.test_no_label) ... skipped "label whitelist
{'base'}"

----------------------------------------------------------------------
Ran 3 tests in 0.001s

OK (skipped=2)
```

因为在第 1 条测试用例中设置了"base"标签，所以"base"标签中的测试用例被执行，其他测试用例则被跳过。

- 设置黑名单为"slow"。

```
> python test_label.py  # 黑名单
```

```
test_label_base (__main__.MyTest.test_label_base) ... ok
test_label_slow (__main__.MyTest.test_label_slow) ... skipped "label blacklist
{'slow'}"
test_no_label (__main__.MyTest.test_no_label) ... ok

----------------------------------------------------------------------
Ran 3 tests in 0.001s

OK (skipped=1)
```

因为在第 2 条测试用例中设置了"slow"标签，所以"slow"标签中的测试用例被跳过，其他测试用例被执行。

7.3 使用缓存

在自动化测试过程中，经常需要通过缓存临时记录一些数据，以减少不必要的操作。比如，由于多条测试用例会公用登录 token，因此可以借助缓存来暂存登录 token 的数据，以减少在测试过程中重复登录的时间消耗。

实现缓存的方式有多种：

- Redis：Redis 是一款开源的基于内存的键值对存储系统。
- LRU Cache：基于 LRU 算法实现的缓存技术。
- 磁盘文件模拟 Cache：缓存的核心是存储、读写数据，因此我们可以利用磁盘文件读写模拟缓存。

7.3.1 Redis 的使用

Redis 的全称为 Remote Dictionary Server（远程数据服务），其主要被当作高性能缓存服务器使用，当然也可以作为消息中间件和 Session 共享等。

Redis 广泛应用于各种系统，通过它可以减少对数据库的访问，从而提高系统数据的访问速度。

Redis 的安装方式有很多种，对于 Linux 或 macOS 操作系统，可以直接使用包管理工具安装。

```
> apt install redis    # Ubuntu
> yum install redis    # Cent OS
> brew install redis   # macOS
```

下面通过 redis-server 启动 Redis 服务。

```
> redis-server
92052:C 31 Oct 2023 16:31:13.564 # oO0OoO0OoO0Oo Redis is starting oO0OoO0OoO0Oo
92052:C 31 Oct 2023 16:31:13.564 # Redis version=7.0.11, bits=64, commit=000000
```

```
00, modified=0, pid=92052, just started
92052:C 31 Oct 2023 16:31:13.564 # Warning: no config file specified, using the
 default config. In order to specify a config file use redis-server /path/to/re
dis.conf
92052:M 31 Oct 2023 16:31:13.565 * Increased maximum number of open files to 10
032 (it was originally set to 2560).
92052:M 31 Oct 2023 16:31:13.565 * monotonic clock: POSIX clock_gettime
```

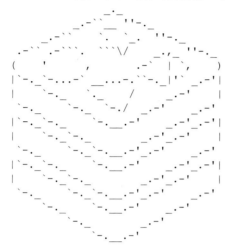

```
                          Redis 7.0.11 (00000000/0) 64 bit

                          Running in standalone mode
                          Port: 6379
                          PID: 92052

                               https://red***.io
```

```
92052:M 31 Oct 2023 16:31:13.566 # WARNING: The TCP backlog setting of 511 cann
ot be enforced because kern.ipc.somaxconn is set to the lower value of 128.
92052:M 31 Oct 2023 16:31:13.566 # Server initialized
92052:M 31 Oct 2023 16:31:13.566 * Ready to accept connections
```

用 pip 命令安装 redis-py，即 Redis 的 Python 客户端库。

> pip install redis

使用示例

在自动化测试中使用 Redis。

```
import hashlib
import unittest
from time import time
import redis


redis_client = redis.Redis(host='localhost', port=6379)

def login(username: str, password: str) -> any:
    """
    模拟登录，生成token
    """
    global redis_client
    token = redis_client.get(username)
    # 如果有token，则直接返回
```

```python
        if token is not None:
            return token

        src = f'{username}.{password}'
        m = hashlib.md5()
        m.update(src.encode("utf-8"))
        token = m.hexdigest()
        # 写入 token
        redis_client.set(username, token)
        return token

class CacheTest(unittest.TestCase):

    def setUp(self):
        self.start = time()

    def tearDown(self):
        end = time()
        print(f"run time:{end - self.start}")

    def test_case1(self):
        token = login("admin", "abc123")
        print(f"case1 token: {token}")

    def test_case2(self):
        token = login("admin", "abc123")
        print(f"case2 token: {token}")

    def test_case3(self):
        token = login("admin", "abc123")
        print(f"case3 token: {token}")


if __name__ == '__main__':
    unittest.main()
```

📖 **代码说明**

导入 redis 模块，接着使用 Redis 类创建连接，默认 host 为本机 localhost，端口号为 6379。

实现 login() 方法，先通过 get() 方法获取 key 对应的 value 信息。如果有，则直接返回；如果没有，则用 username+password 变量值生成 md5 值。

实现三条测试用例，分别调用 login() 方法传入相同的参数。第一条测试用例在生成 token 之后，会将 token 保存到 Redis 服务器中，后面两条测试用例则直接返回 Redis 服务器中保存的结果。

运行结果

```
> python test_redis.py
case1 token: b'fccd7ab1a9d96d2fae0f7c37ce4abe01'
run time:2.0560872554779053
.test_case2 token: b'fccd7ab1a9d96d2fae0f7c37ce4abe01'
run time:0.0
.test_case3 token: b'fccd7ab1a9d96d2fae0f7c37ce4abe01'
run time:0.0
.
----------------------------------------------------------------------
Ran 3 tests in 2.048s

OK
```

从运行结果可以看出,我们分别记录了每条测试用例的运行时间,第一条测试用例建立 Redis 连接并读写数据需要 2s,后面两条测试用例耗时基本可以忽略不计。

7.3.2　LRU Cache

LRU 的全称是 Least Recently Used,其中文含义是"最近最久未使用"。LRU 算法的设计原则是:如果一个数据在最近一段时间内没有被访问,那么将来它被访问的可能性也很小。

Python 提供了 lru_cache 装饰器,其用法非常简单,只装饰需要缓存的方法即可。

使用示例

```python
import hashlib
import unittest
from time import time
from functools import lru_cache

@lru_cache(maxsize=None)
def login(username: str, password: str) -> any:
    """
    模拟登录,生成token
    """
    src = f'{username}.{password}'
    m = hashlib.md5()
    m.update(src.encode("utf-8"))
    token = m.hexdigest()
    return token

class CacheTest(unittest.TestCase):

    def setUp(self):
        self.start = time()

    def tearDown(self):
```

```python
        end = time()
        print(f"run time:{end - self.start}")

    def test_case1(self):
        token = login("admin", "abc123")
        print(f"case1 token: {token}")

    def test_case2(self):
        token = login("admin", "abc123")
        print(f"case2 token: {token}")

    def test_case3(self):
        token = login("admin", "abc123")
        print(f"case3 token: {token}")

if __name__ == '__main__':
    unittest.main()
```

📖 **代码说明**

这里使用了 lru_cache 装饰器的 login() 方法，第一次执行 login() 方法时会缓存返回的结果。第二次执行 login() 方法时会先判断参数，如果与上一次的不同，则重新执行 login() 方法。如果与上一次的相同，则直接从缓存中获取上次的结果并返回。

⌛ **运行结果**

```
> python test_lru_cache.py
case1 token: fccd7ab1a9d96d2fae0f7c37ce4abe01
run time:0.0009856224060058594
.case2 token: fccd7ab1a9d96d2fae0f7c37ce4abe01
run time:0.0
.case3 token: fccd7ab1a9d96d2fae0f7c37ce4abe01
run time:0.0
.
----------------------------------------------------------------------
Ran 3 tests in 0.001s

OK
```

从运行结果来看，除 login() 方法本身的执行时间外，lru_cache 装饰器在写缓存时基本不耗时。因为它和 Redis 一样，也是基于内存的一种缓存技术。

7.3.3 磁盘文件模拟 Cache

前面两种缓存方式都是基于内存的技术方案，除此之外，我们可以基于磁盘文件的读写操作，实现数据的缓存。因为是基于磁盘文件的读写，所以数据可以持久保存。

📁 目录结构

```
07_chapter/
├──extends
│   ├──cache.py
│   ├──cache_data.json
│   └── __init__.py
└──test_cache.py
```

创建 cache.py 文件，实现数据的读、写、清除等相关 API。

```python
import os
import json

FILE_DIR = os.path.dirname(os.path.abspath(__file__))
DATA_PATH = os.path.join(FILE_DIR, "cache_data.json")


class Cache:
    """
    读写 JSON 文件，实现 Cache 类
    """

    def __init__(self):
        """
        初始化 Cache 文件
        """
        is_exist = os.path.isfile(DATA_PATH)
        if is_exist is False:
            with open(DATA_PATH, "w", encoding="utf-8") as json_file:
                json.dump({}, json_file)

    @staticmethod
    def clear(key: str = None) -> None:
        """
        清除 Cache 数据
        """
        if key is None:
            with open(DATA_PATH, "w", encoding="utf-8") as json_file:
                print("Clear all cache data")
                json.dump({}, json_file)
        else:
            with open(DATA_PATH, "r+", encoding="utf-8") as json_file:
                save_data = json.load(json_file)
                del save_data[key]
                print(f"Clear cache data: {key}")

            with open(DATA_PATH, "w+", encoding="utf-8") as json_file:
                json.dump(save_data, json_file)

    @staticmethod
```

```python
    def set(data: dict) -> None:
        """
        添加 Cache 数据
        """
        with open(DATA_PATH, "r+", encoding="utf-8") as json_file:
            save_data = json.load(json_file)
            for key, value in data.items():
                data = save_data.get(key, None)
                if data is None:
                    print(f"Set cache data: {key} = {value}")
                else:
                    print(f"update cache data: {key} = {value}")
                save_data[key] = value

        with open(DATA_PATH, "w+", encoding="utf-8") as json_file:
            json.dump(save_data, json_file)

    @staticmethod
    def get(name=None):
        """
        获取 Cache 数据
        """
        with open(DATA_PATH, "r+", encoding="utf-8") as json_file:
            save_data = json.load(json_file)
            if name is None:
                return save_data

            value = save_data.get(name, None)
            if value is not None:
                print(f"Get cache data: {name} = {value}")
            return value


cache = Cache()
```

> 📖 **代码说明**

首先，实现 Cache 类，在 __init__() 初始化方法中判断 JSON 文件是否存在。如果不存在，则创建 JSON 文件。

其次，分别实现三个方法。

- clear()：清除 Cache 数据，默认清除整个 JSON 文件的数据。如果指定了字典的 key，则仅删除这一项数据。
- set()：添加 Cache 数据，并以 dict 格式保存。首先，读取数据，判断 key 是否存在。如果 key 存在，则更新现有数据的 value。如果 key 不存在，则新增选项，重新保存数据。
- get()：获取 Cache 数据，默认获取整个 JSON 文件的数据。也可以指定 key，但仅获取某项数据。

使用示例

```python
import hashlib
import unittest
from time import time
from extends.cache import cache

def login(username: str, password: str) -> any:
    """
    模拟登录,生成token
    """
    src = f'{username}.{password}'
    m = hashlib.md5()
    m.update(src.encode("utf-8"))
    token = m.hexdigest()
    token_value = cache.get("token")
    if token_value is None:
        cache.set({"token": token})
    return token

class CacheTest(unittest.TestCase):

    @classmethod
    def setUpClass(cls):
        # 清空所有缓存
        cache.clear()

    @classmethod
    def tearDownClass(cls):
        # 获取所有缓存
        all_token = cache.get()
        print(f"all: {all_token}")

    def setUp(self):
        self.start = time()

    def tearDown(self):
        end = time()
        print(f"run time:{end - self.start}")

    def test_case1(self):
        token = login("admin", "abc123")
        print(f"case1 token: {token}")

    def test_case2(self):
        token = login("admin", "abc123")
        print(f"case2 token: {token}")
```

```python
    def test_case3(self):
        token = login("admin", "abc123")
        print(f"case3 token: {token}")


if __name__ == '__main__':
    unittest.main()
```

Cache 类的用法比较简单，setUpClass()函数会清空所有缓存，tearDownClass()函数会获取所有缓存，其他内容与前面的示例类似。

⌛ 运行结果

```
> python test_cache.py
Clear all cache data
Set cache data: token = fccd7ab1a9d96d2fae0f7c37ce4abe01
case1 token: fccd7ab1a9d96d2fae0f7c37ce4abe01
run time:0.0
.Get cache data: token = fccd7ab1a9d96d2fae0f7c37ce4abe01
case2 token: fccd7ab1a9d96d2fae0f7c37ce4abe01
run time:0.0
.Get cache data: token = fccd7ab1a9d96d2fae0f7c37ce4abe01
case3 token: fccd7ab1a9d96d2fae0f7c37ce4abe01
run time:0.0009899139404296875
.all: {'token': 'fccd7ab1a9d96d2fae0f7c37ce4abe01'}
----------------------------------------------------------------------
Ran 3 tests in 0.003s

OK
```

本节介绍了三种缓存的实现，每种方式都有优缺点，三种缓存的对比如表 7-1 所示。

表 7-1 三种缓存的对比

对比项	名称		
	Redis	LRU Cache	磁盘文件模拟Cache
读写位置和速度	内存，快	内存，快	磁盘，一般
持久化	计算机重启，数据丢失	计算机重启，数据丢失	计算机重启，数据仍然保留
使用方式	一般，需要手动判断数据是否存在	简单，通过装饰器判断数据是否存在	一般，需要手动判断数据是否存在
外部依赖	需要依赖Redis服务器	无外部依赖	无外部依赖

7.4 实现日志

无论是写测试代码还是写业务代码，日志都可以帮助定位问题。

记录日志最简单的方法就是在你想要记录的地方使用 print()方法。在简单的代码中或者小型项目中这样做是没有问题的，但是在一些较大的项目中，有时候定位问题需要查看日志，此时用 print()方法就不适合了。

print 语句默认打印的日志既没有日期时间、报错文件名、行号和日志级别等，也无法把日志输出到指定文件中。而这些都是日志默认会提供的功能，因此直接使用日志即可。

在 Python 中使用日志的方式有两种。
- 第 1 种，使用 Python 自带的 logging 模块。
- 第 2 种，使用 Python 的第三方日志库，比如，StructLog 库和 Loguru 库。

下面介绍具体用法。

7.4.1 logging 模块

📂 目录结构

```
07_chapter/
├──lib_demo
│  └──logging_demo.py
├──log
│  └──my_log.log
├──extends
│  ├──my_logging.py
│  └── __init__.py
└──test_my_logging.py
```

🔖 使用示例

logging 模块的基本用法如下。

```
import logging

logging.basicConfig(level=logging.DEBUG)

logging.debug("this is a debug")
logging.info("this is a info")
logging.warning("this is a warning")
logging.error("this is an error")
logging.critical("this is a critical")
```

📖 代码说明

basicConfig()方法用于设置 logging 模块的基本配置，level 参数用于指定日志的打印级别。

日志共分为五个等级：DEBUG、INFO、WARNING、ERROR 和 CRITICAL。

运行结果

```
> python logging_demo.py
DEBUG:root:this is a debug
INFO:root:this is a info
WARNING:root:this is a warning
ERROR:root:this is an error
CRITICAL:root:this is a critical
```

上面介绍的日志记录，其实是通过一个被称作日志记录器（Logger）的实例对象创建的。每个记录器都有一个名称，当直接使用 logging 模块记录日志时，系统会默认创建名为 root 的记录器，这个记录器就是根记录器。

创建 extends/my_logging.py 文件，实现自定义 MyLog 类。

```python
# extends/my_logging.py
import logging
from logging import FileHandler
from logging import StreamHandler

class MyLog:

    def __init__(self, name=__name__, level=logging.DEBUG, logfile="log.log"):
        """
        日志基本配置
        :param name:    模块名称
        :param level:   日志级别
        :param logfile: 日志文件
        """

        # 自定义日志格式
        formatter = logging.Formatter('[%(asctime)s] %(name)s - %(levelname)s - %(message)s')

        self.logger = logging.getLogger(name)
        self.logger.setLevel(level)

        # 输出流
        stream_handler = StreamHandler()
        stream_handler.setFormatter(formatter)
        self.logger.addHandler(stream_handler)

        # 文件输出
        file_handler = FileHandler(filename=logfile)
        file_handler.setFormatter(formatter)
        self.logger.addHandler(file_handler)

    def debug(self, message: str) -> None:
        """ debug log """
```

```python
        self.logger.debug(message)

    def info(self, message: str) -> None:
        """ info log """
        self.logger.info(message)

    def warning(self, message: str) -> None:
        """ warning log """
        self.logger.warning(message)

    def error(self, message: str) -> None:
        """ error log """
        self.logger.error(message)

    def critical(self, message: str) -> None:
        """ critical log """
        self.logger.critical(message)


if __name__ == '__main__':
    log = MyLog()
    log.logger.debug("this is a debug")
```

📖 代码说明

Formatter 类用于定义自定义日志格式。通过设置 setFormatter() 方法可使自定义日志格式生效。

getLogger 类即日志记录器，日志记录器的名称可以是任意名称，不过最佳实践是直接把模块的名称（__name__）当作日志记录器的名称。

StreamHandler 类定义了输出流，可以理解为控制台输出。

FileHandler 类定义了文件输出，同样可以将日志保存到日志文件中。在服务器上这一功能尤为重要，因为通过日志文件定位问题是线上排查问题的重要方式。

最终通过 addHandler() 方法添加 handler，使配置生效。

🖊 使用示例

调用 my_logging 文件封装 MyLog 类。

```python
import logging
from extends.my_logging import MyLog

log = MyLog(__name__, level=logging.DEBUG, logfile="./log/my_log.log")

log.debug("this is a debug")
log.info("this is a info")
log.warning("this is a warning")
```

```
log.error("this is an error")
log.critical("this is a critical")
```

⌛ 运行结果

终端日志输出如下。

```
> python test_my_logging.py

[2023-06-24 21:47:48,536] __main__ - DEBUG - this is a debug
[2023-06-24 21:47:48,537] __main__ - INFO - this is a info
[2023-06-24 21:47:48,537] __main__ - WARNING - this is a warning
[2023-06-24 21:47:48,537] __main__ - ERROR - this is an error
[2023-06-24 21:47:48,537] __main__ - CRITICAL - this is a critical
```

log 目录中的 my_log.log 文件输出如下。

```
# my_log.log
[2023-06-24 21:47:48,536] __main__ - DEBUG - this is a debug
[2023-06-24 21:47:48,537] __main__ - INFO - this is a info
[2023-06-24 21:47:48,537] __main__ - WARNING - this a is warning
[2023-06-24 21:47:48,537] __main__ - ERROR - this is an error
[2023-06-24 21:47:48,537] __main__ - CRITICAL - this is a critical
```

7.4.2 Loguru 库

Loguru 库可以为 Python 带来令人愉快的日志记录。该库通过添加一系列有用的功能解决了标准记录器的注意事项，从而使 Python 日志记录的使用变得不再痛苦。在应用程序中使用日志应该是一种自动操作，Loguru 库试图使其既令人愉快又强大。

用 pip 命令安装 Loguru 库。

```
> pip install loguru
```

💡 使用示例

Loguru 库是预先配置的，会先输出到 stderr。如果没有特殊要求，则 Loguru 库可以开箱即用。

```
# loguru_demo.py
from loguru import logger

logger.debug("this is a debug!")
logger.info("this is a info!")
logger.warning("this is a warning!")
logger.error("this is an error!")
logger.success("this is a success!")
```

⌛ 运行结果

```
> python loguru_demo.py
```

```
2023-06-24 23:30:21.725 | DEBUG   | __main__:<module>:3 - this is a debug!
2023-06-24 23:30:21.727 | INFO    | __main__:<module>:4 - this is a info!
2023-06-24 23:30:21.728 | WARNING | __main__:<module>:5 - this is a warning!
2023-06-24 23:30:21.729 | ERROR   | __main__:<module>:6 - this is an error!
2023-06-24 23:30:21.731 | SUCCESS | __main__:<module>:7 - this is a success!
```

Loguru 库同样支持自定义日志格式，配置日志级别和日志文件等。

```python
# loguru_demo2.py
import sys
import time
from loguru import logger

# 删除默认配置
logger.remove()

# 自定义日志格式
log_format = "<green>{time:YYYY-MM-DD HH:mm:ss}</> {file} <level>| {level} | {message}</level>"

# 添加控制台打印
logger.add(sys.stderr, format=log_format, level="DEBUG")

# 添加日志文件打印
logger.add(f"file_{time.time()}.log", rotation="12:00", format=log_format)

logger.debug("this is a debug!")
logger.info("this is a info!")
logger.warning("this is a warning!")
logger.error("this is an error!")
logger.success("this is a success!")
```

📖 代码说明

remove()方法可以删除默认配置。log_format 变量可以自定义日志格式。

add()方法用于添加日志配置。第一个 add()方法添加控制台打印，第二个 add()方法添加日志文件打印。format 参数用于指定日志格式，level 参数定义了日志级别，而 rotation 参数设置了以日志记录的大小或时间等方式对日志文件进行分割。

⏳ 运行结果

- 终端输出如下。

```
> python loguru_demo2.py

2023-06-24 23:32:51 loguru_demo2.py | DEBUG | this is a debug!
2023-06-24 23:32:51 loguru_demo2.py | INFO | this is a info!
2023-06-24 23:32:51 loguru_demo2.py | WARNING | this is a warning!
```

```
2023-06-24 23:32:51 loguru_demo2.py | ERROR   | this is an error!
2023-06-24 23:32:51 loguru_demo2.py | SUCCESS | this is a success!
```

- 日志文件输出（file_1672642489.2261276.log）如下。

```
2023-06-24 23:32:51 loguru_demo2.py | DEBUG   | this is a debug!
2023-06-24 23:32:51 loguru_demo2.py | INFO    | this is a info!
2023-06-24 23:32:51 loguru_demo2.py | WARNING | this is a warning!
2023-06-24 23:32:51 loguru_demo2.py | ERROR   | this is an error!
2023-06-24 23:32:51 loguru_demo2.py | SUCCESS | this is a success!
```

7.5 自定义异常

如果你要开发的是一个库或框架，那么自定义异常有很多好处。比如，可以清楚地显示潜在的错误，让函数和模块更具可维护性。我们日常在使用 Selenium 时，经常看到 NoSuchElementException、NoSuchAttributeException 等异常类。通过这些异常类，我们可以快速地将错误锁定到 Selenium 的"元素"或"属性"上。

📂 目录结构

```
07_chapter/
├──extends
│   ├──exceptions.py
│   └── __init__.py
└──test_exceptions.py
```

在 Python 中实现自定义异常类。

```python
# exception.py

class CustomException(Exception):
    """
    自定义异常类
    """

    def __init__(self, msg: str = None):
        self.msg = msg

    def __str__(self):
        exception_msg = f"Message: {self.msg}\n"
        return exception_msg


class AgeError(CustomException):
    """
    Age error
    """
    pass
```

```python
class HeightError(CustomException):
    """
    Height error
    """
    pass


class WeightError(CustomException):
    """
    Weight error
    """
    pass
```

📖 **代码说明**

CustomException 类继承自 Exception 异常类，msg 参数定义了具体的异常信息。AgeError、HeightError 和 WeightError 可以根据需求为它们定义不同的异常类。

🔖 **使用示例**

下面通过例子演示如何在具体的场景中使用自定义异常类。

```python
from extends.exception import AgeError, HeightError, WeightError

class_student = {
    "小红": {"age": 11, "height": 135,"weight": 54},
    "小明": {"age": 12, "height": 142,"weight": 60},
    "小刚": {"age": 12, "height": 143,"weight": 200}
}


class Student:

    def __init__(self, name):
        self.name = name

    def physical_examination(self):
        student = class_student[self.name]
        if student["age"] > 20:
            raise AgeError("age 超标")
        if student["height"] > 180:
            raise HeightError("height 超标")
        if student["weight"] > 120:
            raise WeightError("weight 超标")

if __name__ == '__main__':
    s = Student("小刚")
    grade = s.physical_examination()
```

```
        print(grade)
```

📖 **代码说明**

Student 类通过 name 接收一个学生名字，之后通过学生名字获取该学生的体检指标。我们对学生的年龄、身高、体重设置了上限，一旦某项指标大于上限，就抛出异常。

⏳ **运行结果**

```
> python .\test_exception.py

Traceback (most recent call last):
  File "D:\github\test-framework-design\07_chapter\test_exception.py", line 27, in <module>
    grade = s.physical_examination()
            ^^^^^^^^^^^^^^^^^^^^^^^^
  File "D:\github\test-framework-design\07_chapter\test_exception.py", line 22, in physical_examination
    raise WeightError("weight 超标")
extends.exception.WeightError: Message: weight 超标
```

最后，我们需要理解异常和日志的区别。异常用于程序中断，也就是说，如果没有按照预设范围传参，程序将终止执行；而日志更多的是提供打印信息，帮助你了解程序的运行过程。

第 8 章 Web UI 自动化测试设计

十几年前,在笔者刚进入软件测试行业时,HP 公司的自动化测试工具 QTP(Quick Test Professional)是行业的主流,后来开源自动化工具 Selenium 逐渐流行。在此后的数年间,Selenium 得到了广泛的应用,几乎成为 Web UI 自动化测试的必备工具。最近几年,在 Web UI 自动化测试领域出现了一些新的自动化测试工具,如 Cypress、Playwright,大有与 Selenium 三足鼎立之势。

8.1 主流 Web 测试库

8.1.1 Selenium

目前,Selenium 大版本已经更新到了 4.0 版,虽然仍然在稳步迭代中,但并没有太多突破性更新。Selenium 在以下方面有着明显的优势。

(1)支持更多的语言和浏览器。

Selenium 官方支持 Java、Python、Ruby、C#、C++、JavaScript、Rust 等编程语言,支持 Chrome、Chromium、Edge、Firefox、IE、Safari 等浏览器,为想要使用 Selenium 的工程师提供了便利,你只需使用你熟悉的编程语言就可以使用 Selenium。

(2)Selenium 支持分布式运行。

Selenium Grid 可以轻松地在多台主机上实现分布式运行。利用 Docker 技术,docker-selenium 可以轻松启动多个 Selenium 运行环境。

下面以 Python 为例,介绍 Selenium 的使用。

用 pip 命令安装 Selenium 库。

```
> pip install selenium
```

安装 Chrome 或 Firefox 浏览器。在 Selenium 4.6 版本之后，无须手动安装和管理浏览器驱动。

❥ **使用示例**

编写自动化测试用例。

```
# selenium_demo.py
from selenium import webdriver

# 安装Chrome或Firefox浏览器
# browser = webdriver.Chrome()
browser = webdriver.Firefox()

browser.get('http://selen***.dev/')

assert browser.title == "Selenium"
```

8.1.2 Cypress

使用 Cypress，你可以轻松地为现代 Web 应用程序创建测试，可视化地调试它们，并在持续集成构建中自动运行它们。

在 Vue.js 项目中，笔者使用了 Cypress 进行前端 UI 自动化测试。由于 Cypress 默认使用 JavaScript 作为编程语言，可以更好地和前端项目进行整合，因此提供了极佳的用户体验！

（1）集成浏览器及驱动。

仅需一条命令即可安装 Cypress，之后所有必要的环境配置都将自动完成。与 Selenium 相比，Cypress 无须单独安装和管理浏览器及驱动。

（2）运行速度更快。

在运行测试之前，Cypress 会使用 webpack 将测试代码中的所有模块都放入一个 .js 文件中，再启动浏览器，并将测试代码注入一个空白页面，然后在浏览器中运行。与 Selenium 相比，Cypress 无须每次运行都启动浏览器。

Cypress 在测试代码和应用程序时，均运行在由 Cypress 全权控制的浏览器中，且它们运行在同一个 Domain 下的不同 iframe 内，所以 Cypress 的测试代码可以直接操作 DOM、Windows 对象和 localStorage，而无须通过网络访问，这就是 Cypress 运行速度更快的原因。

（3）支持更多类型的测试。

- E2E 测试：即 End-to-End Testing（端到端测试），我们也可以认为是 UI 测试。
- 组件测试：即 Component Testing。一个 Web 页面由许多组件组成，比如，按钮、输入框、下拉框和表格。Cypress 可以针对单个组件进行测试。
- 集成测试：即 Integration Testing，笔者认为这里的集成测试为接口测试（或 API 测试），Cypress 集成了 HTTP 接口，可以进行 HTTP 接口测试。
- 单元测试：即 Unit Testing，Cypress 仅支持用 JavaScript 或 TypeScript 语言编写的测试脚本，因此这里的单元测试功能也仅仅针对 JavaScript 或 TypeScript 语言。

Cypress 目前不支持多语言环境，只能基于 Node.js 运行。

首先按照 Node 官方网站提供的方法安装 Node，这里不再赘述，然后按照下面的命令安装 Cypress。

```
> npm install cypress --save-dev
```

启动 Cypress。

```
> npx cypress open
```

在 Cypress 窗口中选择测试类型：E2E Testing 或 Component Testing，如图 8-1 所示。

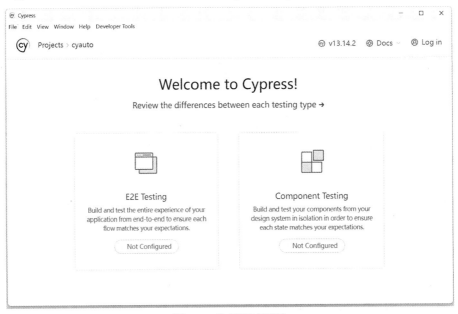

图 8-1　选择测试类型

单击"E2E Testing"选项，进入浏览器选择页面，如图 8-2 所示。

第 8 章　Web UI 自动化测试设计　　129

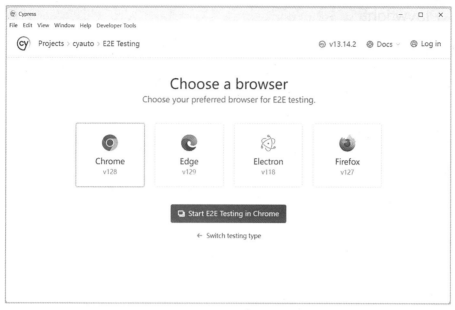

图 8-2　浏览器选择页面

选择"Chrome"项，Cypress 默认提供了一些示例，我们可以选择"todos"实例并执行，如图 8-3 所示。

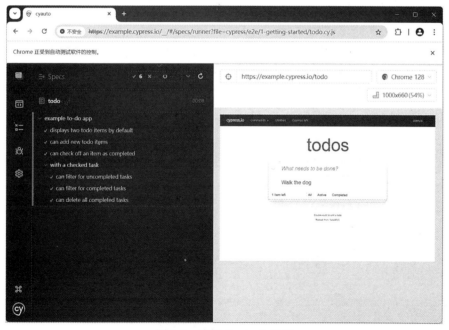

图 8-3　选择"todos"实例

8.1.3 Playwright

Playwright 是由微软推出的一款自动化测试工具，它提供了更好的开发、调试体验。由于它支持多种编程语言和浏览器，所以正在被越来越多的开发工程师和测试工程师使用。未来它很有可能替代 Selenium，成为测试工程师的首选。

（1）集成浏览器及驱动。

与 Cypress 相同，Playwright 直接集成了自动化测试运行环境，无须单独安装浏览器及驱动。

（2）支持多语言。

Playwright 支持 TypeScript、JavaScript、Java、Python 和 .NET 等语言，开发工程师和测试工程师可以根据自己熟悉的编程语言使用 Playwright。

（3）更好地开发、调试、运行体验。

- 开发方面：提供了录制功能；API 支持断言，可以更方便地实现多窗口；支持同步和异步两种写法。
- 调试方面：Playwright Inspector 支持运行和调试测试用例。
- 运行方面：与 Selenium 相比，运行速度更快。

下面以 Python 为例，介绍 Playwright 的用法。

用 pip 命令安装 Playwright。

```
> pip install playwright
```

用 playwright 命令安装浏览器。

```
> playwright install    # 默认安装所有浏览器
> playwright install --force chromium  # 安装指定浏览器
```

运行 Codegen。

```
> playwright codegen bing.com
```

录制 bing 测试脚本，搜索 playwright 关键字，并跳转到 Playwright 中文网，如图 8-4 所示。

第 8 章　Web UI 自动化测试设计　131

图 8-4　录制 bing 测试脚本

停止录制，复制录制的脚本，将其保存为 playwright_demo.py 文件。

```
# playwright_demo.py
from playwright.sync_api import Playwright, sync_playwright, expect


def run(playwright: Playwright) -> None:
    browser = playwright.chromium.launch(headless=False)
    context = browser.new_context()
    page = context.new_page()
    page.goto("https://cn.bi***.com/")
    page.get_by_role("searchbox", name="输入搜索词").click()
    page.get_by_role("searchbox", name="输入搜索词").fill("playwright")
    page.get_by_role("searchbox", name="输入搜索词").press("Enter")
    with page.expect_popup() as page1_info:
        page.get_by_role("link", name="Playwright 中文网", exact=True).click()
    page1 = page1_info.value
    page1.get_by_text("Playwright", exact=True).click()

    # ---------------------
    context.close()
    browser.close()


with sync_playwright() as playwright:
    run(playwright)
```

8.2 Selenium API 的二次开发

Selenium API 在使用过程中往往需要进行二次开发，一般从以下几个方面对 Selenium API 进行再次封装。

- 简化方法名称：Selenium 中的部分方法名称较长。比如，定位方法的名称是 find_element(By.PARTIAL_LINK_TEXT, "xx")，此时可以简化方法名称。
- 智能等待：Selenium 提供了元素定位和等待 API。在编写元素定位时往往需要手动设置等待，显然这比较麻烦，在二次开发时会重点对元素定位和等待进行整合。
- API 整合：很多时候，有些动作会组合使用。比如，对于输入框，一般会先清空输入框，再在输入框中输入内容，最后按回车键，因此可以将 clear()、send_keys() 和 Keys.ENTER 等 API 组合成一个方法使用。
- 方法调用方式变化：比如，"方法链"。

下面将对基于 Selenium 的一些主流自动化测试项目进行归类和总结，看看可以通过哪些方式对 Selenium API 进行二次开发。

原生的 Selenium 代码如下。

```
from selenium import webdriver
from selenium.webdriver.common.by import By

dr = webdriver.Chrome()
dr.get("https://www.bai**.com")

dr.find_element(By.ID, "kw").send_keys("selenium")
dr.find_element(By.ID, "su").click()

dr.quit()
```

代码非常简单，启动 Chrome 浏览器，访问百度首页，在搜索框中输入"selenium"关键字，单击搜索按钮（或按回车键触发搜索）。

📂 目录结构

```
08_chapter/
├──selenium_api
│   ├──se_api_one.py
│   ├──se_api_two.py
│   ├──se_api_three.py
│   ├──se_api_four.py
│   └──__init__.py
├──test_se_api_one.py
├──test_se_api_two.py
├──test_se_api_three.py
└──test_se_api_four.py
```

8.2.1 封装：重命名方法

重命名方法主要是对 Selenium API 中的方法进行重命名，从而使这些方法更加便于记忆和使用。

```python
# se_api_one.py
from selenium import webdriver
from selenium.webdriver.common.by import By

class Se:

    def __init__(self, timeout: float = 10):
        self.driver = webdriver.Chrome()
        self.driver.implicitly_wait(timeout)

    def open(self, url: str):
        self.driver.get(url)

    def close(self):
        self.driver.close()

    def by_id(self, elem: str):
        return self.driver.find_element(By.ID, elem)

    def by_name(self, elem: str):
        return self.driver.find_element(By.NAME, elem)
```

📖 代码说明

首先调用 Se 类初始化浏览器驱动，然后调用 Se 类下面封装的 by_id()方法和 by_name()方法定位元素对象。open()方法和 close()方法分别用于打开 URL 和关闭浏览器。

🖱 使用示例

```python
from selenium_api.se_api_one import Se

one = Se()
one.open("https://www.bai***.com")
one.by_name("wd").send_keys("selenium")
one.by_id("su").click()
one.close()
```

8.2.2 封装：元素定位和动作整合

元素定位和动作整合主要是将元素定位和操作（或多个操作）整合成一个方法，定位类型和值，并将输入值作为方法的参数。

```python
# se_api_two.py
from selenium import webdriver
```

```python
from selenium.webdriver.common.by import By
from selenium.webdriver.common.keys import Keys

# 定义定位方法对应的字典
LOCATOR_LIST = {
    'id': By.ID,
    'name': By.NAME,
    # ...
}

class Se:

    def __init__(self, timeout: float = 10):
        self.driver = webdriver.Chrome()
        self.driver.implicitly_wait(timeout)

    def open(self, url: str):
        self.driver.get(url)

    def close(self):
        self.driver.close()

    def get_element(self, **kwargs):
        by, value = next(iter(kwargs.items()))
        se_by = LOCATOR_LIST.get(by, None)
        if se_by is None:
            raise NameError("仅支持id或name定位,暂不支持其他定位方式")
        return self.driver.find_element(se_by, value)

    def type(self, text: str, clear: bool = False, enter: bool = False, **kwargs):
        elem = self.get_element(**kwargs)
        if clear:
            elem.send_keys(text)
        elem.send_keys(text)
        if enter:
            elem.send_keys(Keys.ENTER)

    def click(self, **kwargs):
        elem = self.get_element(**kwargs)
        elem.click()
```

📖 **代码说明**

调用 Se 类初始化浏览器驱动,type()和 click()方法既包含元素定位,又包含元素操作。**kwargs 是 keyword Variable Arguments 的缩写,它接收字典形式的参数,这里用于接收元素定位类型和值。通过 get_element()方法对**kwargs 进行解析,可以得到元素对象。

type()方法还包含 clear 参数和 enter 参数。当 clear 参数为 True 时,会对输入框进行清空操作;当 enter 参数为 True 时,在输入文本后会自动回车。

> 使用示例

```
from selenium_api.se_api_two import Se

two = Se()
two.open("https://www.bai***.com")
two.type(name="wd", text="selenium", enter=True)
two.close()
```

注意，name= "wd" 是通过 **kwargs 进行传参的。

8.2.3 封装：独立每个函数

这种封装方式的目的在于将每个操作都独立成一个函数，并通过全局变量 dr 来使用浏览器对象。

```
# se_api_three.py
from selenium import webdriver
from selenium.webdriver.common.by import By

dr = None

def start_chrome(timeout: float = 10):
    global dr
    dr = webdriver.Chrome()
    dr.implicitly_wait(timeout)

def go_to(url):
    global dr
    dr.get(url)

def close():
    global dr
    dr.close()

def id(elem):
    global dr
    return dr.find_element(By.ID, elem)

def name(elem):
    global dr
    return dr.find_element(By.NAME, elem)

def wirte(elem, text):
    elem.send_keys(text)

def click(elem):
    elem.click()
```

📖 代码说明

定义全局变量 dr，每个函数都通过 global dr 使用全局变量。

每个函数都只完成一个动作：id()、name()用于返回定位元素对象，write()、click()用于对元素对象进行输入和单击。

💡 使用示例

```
from selenium_api.se_api_three import *

start_chrome()
go_to("http://www.bai***.com")
write(name("wd"), text="selenium")
click(id("su"))
close()
```

这些方法已经极简了，表面上看各个方法之间没有关联，但其实它们会依赖于 start_chrome() 函数的执行。这个函数用于创建浏览器对象，并赋值给 dr 变量。

8.2.4 封装：链式调用

链式调用，也称为方法链（Method Chaining），就是将一系列的操作、函数和方法像链条一样串起来。

```python
import time
from selenium import webdriver
from selenium.webdriver.common.by import By

# 定义定位方法对应的字典
LOCATOR_LIST = {
    'id': By.ID,
    'name': By.NAME,
    # ...
}

class Se:

    def __init__(self, timeout: float = 10):
        self.driver = webdriver.Chrome()
        self.driver.implicitly_wait(timeout)

    def open(self, url: str):
        self.driver.get(url)
        return self

    def close(self):
        self.driver.close()
        return self
```

```python
def get_element(self, **kwargs):
    by, value = next(iter(kwargs.items()))
    se_by = LOCATOR_LIST.get(by, None)
    if se_by is None:
        raise NameError("仅支持id或name定位,暂不支持其他定位方式")
    return self.driver.find_element(se_by, value)

def sleep(self, sec):
    time.sleep(sec)
    return self

def type(self, text: str, **kwargs):
    elem = self.get_element(**kwargs)
    elem.send_keys(text)
    return self

def click(self, **kwargs):
    elem = self.get_element(**kwargs)
    elem.click()
    return self
```

代码说明

链式调用在代码实现上与"元素定位和动作整合"这个封装类似,唯一的区别在于链式调用会在每个方法执行完毕后都返回 self 实例,这样在调用方法之后就会得到类对象,从而可以继续调用类下面的其他方法。

使用示例

```
from selenium_api.se_api_4 import Se

se = Se()
se.open("https://www.bai***.com").type(name="wd",
text="selenium").click(id="su").close()
```

以上所有示例实现的功能都是一样的:打开浏览器,进入百度页面,输入"selenium",单击"搜索"按钮,关闭浏览器。

与原生 API 相比,以上四种封装方式都较为显著地缩减了代码量,这四种封装方式没有绝对的优劣之分。当然,除代码量外,在封装过程中确保代码的可读性同样至关重要。Selenium 作为一个成熟的 Web UI 自动化工具,有着许多的封装设计。

8.3 Selenium 的断言设计

断言是自动化测试的核心组成部分,通常由测试框架内置提供,其作用是通过验证预期结果与实际结果的一致性，来判定测试用例是否通过或失败。

8.3.1 单元测试框架提供的通用断言

单元测试框架通常仅提供一系列基础的断言方法，用于满足测试中最常见的验证需求。以 unittest 为例，它提供的断言方法如表 8-1 所示。

表 8-1 unittest 提供的断言方法

断言方法	检查对象	引入版本
assertEqual(a, b)	a == b	
assertNotEqual(a, b)	a != b	
assertTrue(x)	bool(x) is True	
assertFalse(x)	bool(x) is False	
assertIs(a, b)	a is b	3.1
assertIsNot(a, b)	a is not b	3.1
assertIsNone(x)	x is None	3.1
assertIsNotNone(x)	x is not None	3.1
assertIn(a, b)	a in b	3.1
assertNotIn(a, b)	a not in b	3.1
assertIsInstance(a, b)	isinstance(a, b)	3.2
assertNotIsInstance(a, b)	not isinstance(a, b)	3.2

这些通用的断言方法并不适用于 Web 自动化测试场景。在 Web 自动化测试实践中，更多的是利用特定于 Web 的断言方法或断言库来验证预期状态，这些通常被称为断言点，比如：

- 断言页面标题；
- 断言页面 URL 地址；
- 断言页面文本信息；
- 断言页面元素；
- 页面截图对比。

以 Selenium 断言页面标题为例。

```
import unittest
from selenium.webdriver import Chrome

class MyTest(unittest.TestCase):

    def test_case(self):
        driver = Chrome()
        driver.get("https://www.selen***.dev/")
```

```python
# 获取当前页面 URL，进行断言
current_title = driver.title
self.assertEqual(current_title, "Selenium")
```

如果进行断言封装，则建议提供一个专门断言页面标题的方法，这样会使代码更加简捷。

```python
self.assertTitle("Selenium")
```

8.3.2　封装 Selenium 断言方法

基于 8.3.1 节的内容，我们除了可以单独封装 Selenium API，还可以基于 unittest 和 Selenium 一并进行封装设计，这样的整合方式，可以提供更加统一的 API。

📁 目录结构

```
08_chapter/
├──common
│   ├──case.py
│   └──__init__.py
└──test_assert.py
```

基于 unittest 增加 Selenium 断言方法。

```python
import unittest
from urllib.parse import unquote
from selenium import webdriver
from selenium.webdriver.common.by import By


# 定义浏览器驱动
class Browser:
    driver = None


# 定义 unittest 主方法
main = unittest.main

# 定位元素的类型
LOCATOR_LIST = {
    'id': By.ID,
    'name': By.NAME,
    'xpath': By.XPATH,
    'css': By.CSS_SELECTOR,
    # ...
}


class TestCase(unittest.TestCase):
    """
    定义 unittest 测试类，实现断言方法
    """

    @classmethod
```

```python
    def setUpClass(cls):
        """
        初始化浏览器驱动
        """
        Browser.driver = webdriver.Chrome()

    @classmethod
    def tearDownClass(cls):
        """
        关闭浏览器
        """
        Browser.driver.close()

    @staticmethod
    def _get_element(**kwargs):
        by, value = next(iter(kwargs.items()))
        se_by = LOCATOR_LIST.get(by, None)
        if se_by is None:
            raise NameError("仅支持通过 id、name、XPath 或 CSS 定位,暂不支持其他定位方式")
        return Browser.driver.find_element(se_by, value)

    @staticmethod
    def open(url):
        """
        打开 URL

        用法:
            self.open(url)
        """
        Browser.driver.get(url)

    def assertTitle(self, title: str = None, msg: str = None) -> None:
        """
        断言当前页面标题是否等于 title

        用法:
            self.assertTitle("title")
        """
        if title is None:
            raise AssertionError("断言的 title 不能为空")

        print(f"assertTitle -> {title}.")
        self.assertEqual(title, Browser.driver.title, msg=msg)

    def assertInTitle(self, title: str = None, msg: str = None) -> None:
        """
        断言当前页面标题是否包含 title

        用法:
```

```python
            self.assertTitle("title")
        """
        if title is None:
            raise AssertionError("断言的 title 不能为空")

        print(f"assertInTitle -> {title}.")
        self.assertIn(title, Browser.driver.title, msg=msg)

    def assertUrl(self, url: str = None, msg: str = None) -> None:
        """
        断言当前页面地址是否为 URL

        用法:
            self.assertUrl("url")
        """
        if url is None:
            raise AssertionError("断言的 URL 不能为空")

        print(f"assertUrl -> {url}.")
        current_url = unquote(Browser.driver.current_url)
        self.assertEqual(url, current_url, msg=msg)

    def assertInText(self, text: str = None, msg: str = None) -> None:
        """
        断言当前页面是否包含 text

        用法:
            self.assertInText("text")
        """
        if text is None:
            raise AssertionError("断言的 text 不能为空")

        elem = Browser.driver.find_element(By.TAG_NAME, "html")
        print(f"assertText -> {text}.")
        self.assertIn(text, elem.text, msg=msg)

    def assertElement(self, **kwargs) -> None:
        """
        断言当前页面定位元素是否存在

        用法:
            self.assertElement(css="#id")
        """
        try:
            self._get_element(**kwargs)
            elem = True
        except Exception as e:
            print("assertElement error", e)
            elem = False
```

```
        print(f"assertElement.")
        self.assertTrue(elem)
```

📖 **代码说明**

定义 Browser 类，其中，driver 变量用于存储浏览器驱动实例。当需要调用 Selenium 提供的浏览器操作时，该类中的所有断言方法，都会基于该 Browser 类实例的 driver 变量来执行。

将 unittest.main 方法赋值给 main，这样做的目的是避免在编写测试用例时导入 unittest 的 API，只需导入 common 的 case 模块即可。

LOCATOR_LIST 常量用于定义定位元素的类型。

创建 TestCase 类，使其继承 unittest 的 TestCase 类，并分别实现以下方法。

- open()：打开浏览器。
- _get_element()：内部方法，返回元素对象。
- assertTitle()：断言标题是否相等。
- assertInTitle()：断言是否包含标题。
- assertUrl()：断言 URL 是否相等。
- assertInText()：断言页面是否包含文本。
- assertElement()：断言页面元素是否存在。

✍ **使用示例**

下面我们通过示例调用前面封装的断言方法。

```
from common import case

class MyTest(case.TestCase):

    def test_case(self):
        self.open("https://www.selen***.dev/")
        self.assertTitle("Selenium")
        self.assertInTitle("Se")
        self.assertUrl("https://www.selen***.dev/")
        self.assertInText("Selenium automates browsers. That's it!")
        self.assertElement(css="#main_navbar > ul > li:nth-child(3) > a")

if __name__ == '__main__':
    case.main()
```

📖 **代码说明**

创建 MyTest 类，使其继承封装的 case 模块的 TestCase 类。

在 test_case 用例中访问 Selenium 官网，调用父类中封装的各个断言方法。通过封装这些断言方法，可以极大地简化 Web 自动化测试中的断言。

8.4 Selenium 环境管理

新的 Web 自动化工具 Cypress、Playwright 在安装的过程中会自动管理浏览器及驱动。早期的 Selenium 是需要手动安装浏览器的，你不仅需要知道浏览器驱动的下载地址，以及与浏览器对应的驱动版本，还需要设置浏览器驱动的环境变量 path。虽然，这些工作是一次性的，但是提高了使用 Selenium 的门槛，好在，Selenium 正在改善这些问题。

8.4.1 Selenium Manager

由于 Selenium 本身一直未提供浏览器驱动管理工具，因此在 Selenium 的生态系统中涌现了多个第三方的浏览器驱动管理项目，如表 8-2 所示。

表 8-2 第三方的浏览器驱动管理项目

语言	驱动管理项目
Java	WebDriverManager
Python	webdriver-manager
JavaScript	webdriver-manager
Ruby	webdrivers
C#	WebDriverManager.Net

Selenium 官方已经意识到这一需求，并旨在提升用户体验，于是在 Selenium 4.6 版本中引入了 Selenium Manager 来解决这一问题。

Selenium Manager 是一个用 Rust 开发的 CLI（命令行界面）工具，允许跨平台执行。当系统在指定路径下未能检测到浏览器驱动程序，或者没有采用第三方驱动程序管理器时，Selenium 会自动启用 Selenium Manager 来进行调用。

Selenium Manager 是随着最新的 Selenium 一起安装的，不需要单独下载。打开终端工具，在任意位置执行"selenium-manager"命令。

```
> selenium-manager --help
selenium-manager 0.4.22
Selenium Manager is a CLI tool that automatically manages the browser/driver
infrastructure required by Selenium.

Usage: selenium-manager.exe [OPTIONS]
Options:
    --browser <BROWSER>
        Browser name (chrome, firefox, edge, iexplorer, safari, or safaritp)
```

```
    --driver <DRIVER>
        Driver name (chromedriver, geckodriver, msedgedriver, IEDriverServer, or
safaridriver)
    --grid [<GRID_VERSION>]
        Selenium Grid. If version is not provided, the latest version is downloaded
    --driver-version <DRIVER_VERSION>
        Driver version (e.g., 106.0.5249.61, 0.31.0, etc.)
    --browser-version <BROWSER_VERSION>
        Major browser version (e.g., 105, 106, etc. Also: beta, dev, canary -or
nightly- is accepted)
    --browser-path <BROWSER_PATH>
        Browser path (absolute) for browser version detection (e.g.,
/usr/bin/google-chrome, "/Applications/Google Chrome.app/Contents/MacOS/Google
Chrome", "C:\Program Files\Google\Chrome\Application\chrome.exe")
...
```

利用"selenium-manager"命令可以方便地管理浏览器、驱动和 Selenium Server 等。

（1）安装浏览器。

```
> selenium-manager --browser chrome
> selenium-manager --browser edge
> selenium-manager --browser firefox
...
```

（2）安装驱动。

```
> selenium-manager --driver chromedriver
> selenium-manager --driver msedgedriver
> selenium-manager --driver geckodriver
...
```

（3）安装 Selenium Server。

```
> selenium-manager --grid 4.22.0
...
```

8.4.2 Docker-Selenium

Docker-Selenium 是 Selenium Grid 服务的 Docker 镜像项目。通过 Docker 创建 Selenium 的运行环境是最简单的，因为省略了浏览器及驱动的安装。

在 Docker 中安装 Firefox 镜像。

```
> docker run -d -p 4444:4444 -p 7900:7900 --shm-size="2g"
selenium/standalone-firefox:4.15.0-20231129
```

用 docker 命令查看镜像。

```
> docker ps | grep "selenium"

a5ad1dd8571a   selenium/standalone-firefox:4.22.0-20240621   "/opt/bin/entry_po
in…"   11 minutes ago   Up 11 minutes   0.0.0.0:4444->4444/tcp, 0.0.0.0:7900->7
900/tcp, 5900/tcp   unruffled_ishizaka
```

查看驱动管理页面。通过浏览器访问 http://localhost:4444，Selenium Grid 页面如图 8-5 所示。

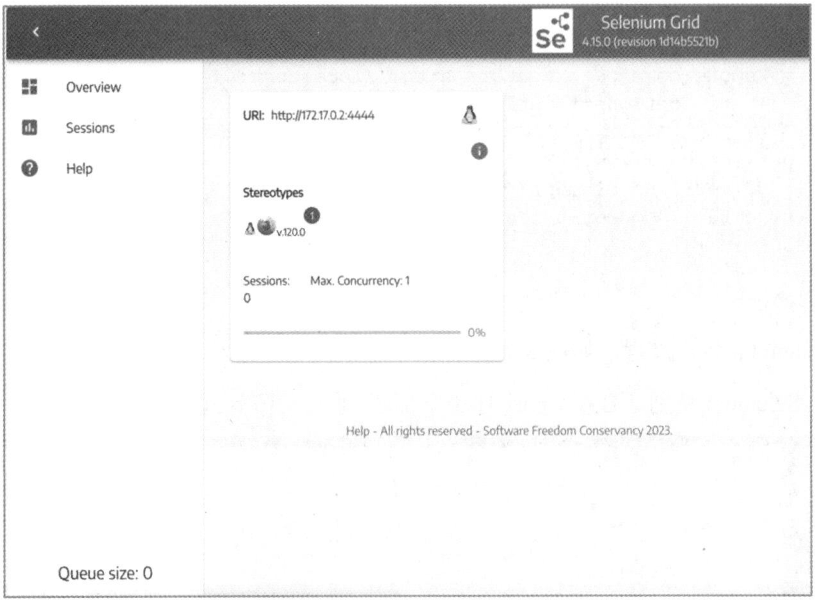

图 8-5 Selenium Grid 页面

通过 NoVNC 可以直观地检查容器活动。通过浏览器访问 http://localhost:7900/?autoconnect=1&resize=scale&password=secret，可以看到如图 8-6 所示页面。

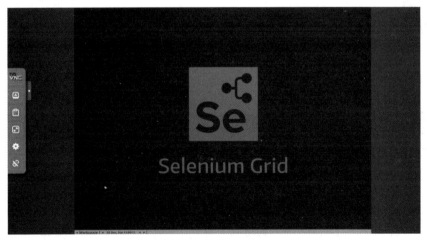

图 8-6 访问 NoVNC

创建 selenium_grid_demo.py 文件，编写 docker-selenium 并使用。

```
# selenium_grid_demo.py
from selenium import webdriver
```

```python
from selenium.webdriver import Remote

# Firefox 浏览器配置
ffOptions = webdriver.FirefoxOptions()
driver = Remote(command_executor='http://localhost:4444/wd/hub',
                options=ffOptions)

driver.get("https://www.b***.com")
search = driver.find_element("id", "sb_form_q")
search.send_keys("docker-selenium")
search.submit()
# ...
driver.close()
```

查看 Selenium 运行过程，如图 8-7 所示。

Docker-Selenium 特别适合在 Linux 环境下部署，尤其是需要远程分布式执行或持续集成时。

图 8-7　Selenium 运行过程

扩展知识

NoVNC 是一种基于 HTML5 技术的远程桌面协议，旨在实现对计算机桌面的远程访问与控制。用户无须在客户端安装额外软件，仅通过标准的 Web 浏览器即可享受接近本地操作系统的体验。得益于 HTML5 技术，NoVNC 能够跨平台运行，兼容多种设备和操作系统环境。它利用 WebSocket 和 HTTP 传输图形化界面数据，以实现远程桌面控制。

第 9 章

App UI 自动化测试设计

App 移动端测试主要涉及 Android 和 iOS 两大平台，还有即将普及的 HarmonyOS Next。在每个平台中又涉及许多测试相关的工具。本章我们聚焦于 Appium 这一成熟且应用广泛的工具。

9.1 App 移动自动化测试工具介绍

App 移动端测试涉及的工具非常多，下面我们对目前主流的移动自动化测试工具进行盘点。

9.1.1 Android 测试工具

AndroidX Test 是一套 Jetpack 库，可以针对 Android 应用进行测试。AndroidX Test 提供了 JUnit4 规则来启动 Activity，并在 JUnit4 测试中与它们进行交互。它还包含 UI 测试框架，比如，Robolectric、Espresso、UI Automator 和 adb。

1. Robolectric

Robolectric 是一个适用于 Android 的快速可靠的单元测试框架。通过 Robolectric，测试可在几秒钟内在工作站的 JVM 上完成运行。

Robolectric 可以模拟 Android 4.1（API 的级别 16）或更高版本的运行时环境，并提供由社区维护的虚假对象（称为"影子"）。通过此功能，你可以测试依赖于框架的代码，而无须使用模拟器或模拟对象。Robolectric 支持 Android 平台的以下几个方面。

- 组件生命周期。
- 事件循环。
- 所有资源。

2. Espresso

Espresso 可以编写简捷、美观且可靠的 Android UI 测试，适用于编写 Android 中型测试或 Android 大型测试。

Espresso 还支持在大型测试中完成以下任务并实现同步。

- 实现跨应用工作流的进程隔离，仅适用于 Android 8.0（API 级别 26）及更高版本。
- 跟踪应用中长时间运行的后台操作。
- 执行设备外测试。

3. UI Automator

UI Automator 是一个 UI 测试框架，特别适用于实现跨应用程序功能测试的场景。它非常适合大规模的测试项目。因为在测试过程中，UI Automator 将 Android 系统和应用程序视为一个黑盒，专注于测试的输入与输出，而非内部实现细节。

UI Automator 的主要特点如下。

- 提供用于检索状态信息并在目标设备上执行操作的 API。
- 支持跨应用 UI 测试的 API。

4. adb

adb（Android Debug Bridge）是一个功能多样的命令行工具，可以让计算机与设备进行通信。它是一个客户端—服务器程序，包括以下三个组件。

- 客户端：用于发送命令。客户端在开发者的机器上运行，用户可以通过在命令行终端中输入 adb 命令来调用客户端。
- 守护进程（adbd）：用于在设备上运行命令。守护程序在每个设备上都作为后台进程持续运行。
- 服务器：用于管理客户端与守护程序之间的通信。服务器在开发者的机器上作为后台进程持续运行。

9.1.2　iOS 测试工具

1. XCTest

XCTest 可用于 iOS 移动应用程序测试，为 Xcode 项目创建并运行单元测试、性能测试和 UI 测试，兼容 Xcode 5.0+。

XCTest 的主要特性如下。

- XCTest 是一个功能强大的 iOS 测试框架，可用于单元测试、性能测试和 UI 测试。

- 无须安装，Xcode 内置了 XCTest，为移动自动化测试提供了启动环境。
- XCTest 提供了对持续集成系统的良好控制能力。
- XCTest 允许用户录制和增强界面测试功能。

2. XCUITest

XCUITest 是一个用于执行 iOS 自动化测试的自动化 UI 测试框架，它被集成在 XCTest 中。

3. iOS 调试工具

iOS 官方没有推出调试工具，下面介绍一些优秀的第三方开源项目。

- Go-iOS 是一个为 iOS 设备提供独立操作系统实现的平台。它可以执行 UI 测试、远程启动或终止应用程序、安装应用程序等操作。它提供了一套操作系统独立的接口，用以实现对 iOS 设备的功能控制。
- idb（iOS Development Bridge）提供了一个灵活的命令行界面，便于用户自动化操控 iOS 模拟器及物理设备。
- sib（sonic ios bridge）是基于 usbmuxd 的 iOS 调试工具。
- tidevice（taobao iOS device）主要用于与 iOS 设备进行通信。

9.1.3 Appium

Appium 是一个开源的工具，同时是一个开源项目及其周边的软件生态系统，专为促进跨多个应用程序平台的 UI 自动化测试而设计，涵盖了移动平台（iOS、Android、Tizen）、网页浏览器（Chrome、Firefox、Safari）、桌面操作系统（macOS、Windows）和电视平台（Roku、tvOS、Android TV、三星）等。

1. Appium

Appium 是基于 W3C WebDriver 协议构建的各种应用程序的跨平台自动化框架。

2. Appium Inspector

Appium Inspector 是由 Appium 提供的移动应用程序的 GUI 检查器，可帮助用户查看 App 元素属性。

3. Client

Appium 支持基于多种语言编写的 Appium 自动化测试脚本，因此，Appium 推出了支持不同编程语言的客户端库，比如 Appium-Python-Client 库、Appium-Java-Client 库、Ruby-Client 库和 AppiumC#-Client 库等。

4. WebDriverAgent

WebDriverAgent 是一个适用于 iOS 的 WebDriver 服务器实现，可用于远程控制 iOS 设备。它通过连接 XCTest.framework，并直接在设备上调用 Apple 的 API 来执行命令。

WebDriverAgent 由 Facebook 开源，目前 Facebook 已停止对该项目的维护。Appium 组织创建了分支并开始维护，WebDriverAgent 在 Appium 组织下得以继续发展。

9.1.4　Airtest Project

Airtest Project 是网易游戏推出的移动自动化测试框架，它提供了一组工具来满足移动自动化测试。

1. Airtest

Airtest 是基于图像识别的跨平台 UI 自动化测试框架，适用于游戏和应用程序，支持的平台有 Windows、Android 和 iOS。

2. Poco

Poco 是一个基于 UI 控件识别的自动化测试框架，目前支持 Unity3D、Cocos2d-x、Android 原生应用、iOS 原生应用和微信 applet 等多个平台。在其他引擎中，你可以通过访问 poco-sdk 来使用 Poco。

3. Airtest IDE

Airtest IDE 是一个跨平台的 UI 自动化测试编辑器，集成了 Airtest 和 Poco 插件，便于用户快速便捷地编写自动化测试脚本。

4. iOS-Tagent

iOS-Tagent 是 WebDriverAgent 项目的一个分支，专门适配于 Airtest，以支持 iOS 移动应用程序的自动化测试。

9.1.5　Open ATX

Open ATX 提供了一组工具来支持移动自动化测试。

1. uiautomator2

uiautomator2 是一个构建在 Android 的 UI Automator 测试框架之上的 Python 测试库。uiautomator2 的原理是在手机上部署一个 HTTP RPC 服务，该服务可以暴露 UI Automator 的核心功能，并通过一套 HTTP 接口进行操作。之后，这些 HTTP 接口被封装成易于使用的 Python 库，从而简化了测试脚本的编写过程。

UI Automator 是由 Google 开发的一个专为 Android 自动化测试设计的用 Java 编写的测试框架，它基于 Accessibility 服务来实现界面元素的交互。

1. facebook-wda

facebook-wda 是一个基于 WebDriverAgent 的 Python 测试库，它通过集成 WebDriverAgent 来实现 iOS 设备的自动化测试功能。

2. adbutils

adbutils 是一个封装了 adb 的 Python 库。adbutils 在 adb 命令的基础上增加了一层封装，使用户可以通过 Python 来调用 adb 的各种命令。

9.2 Appium 基础

App 自动化工具非常多，我们这里选择 Appium 进行重点介绍，主要基于以下几点考虑。
- Appium 是由商业公司维护的，拥有较长的历史，并且仍在持续积极维护中。
- Appium 支持多平台、多语言，并得到了广泛应用。
- Appium 与 Selenium 一脉相承，可以很好地继承 Selenium 的 API 来实现对 App 自动化的支持。

9.2.1 Appium 的安装

目前 Appium 2.0 已经推出稳定版，它带来了一些新的特性。
- 独立的驱动：能够安装和使用基于应用平台的驱动程序（iOS、Android、Windows OS 或 Flutter）。
- 插件生态：通过插件的方式为 Appium 扩展功能。

首先，安装 Node.js。

其次，查看 Node.js 和 npm 版本。Appium 2.0 要求 npm 版本必须在 8.x 以上。

```
> node --version
v18.20.3
> npm --version
10.7.0
```

安装 Appium 2.0。

```
> npm i --location=global appium
```

查看 Appium 版本。

```
> appium --version
```

2.11.1

查看驱动。

```
> appium driver list

✔ Listing available drivers
- espresso [not installed]
- uiautomator2 [not installed]
- xcuitest [not installed]
- mac2 [not installed]
- windows [not installed]
- safari [not installed]
- gecko [not installed]
- chromium [not installed]
```

安装适用于 Android 和 iOS 平台的自动化测试库。

```
> appium driver install espresso
> appium driver install uiautomator2
> appium driver install xcuitest
```

安装 Appium-Python-Client 库。它是 Appium 官方提供的 Python 客户端库。

```
> pip install Appium-Python-Client
```

安装 Appium Inspector。Appium Inspector 是一个适用于移动应用程序的 GUI 检查工具。Appium Inspector 界面如图 9-1 所示。

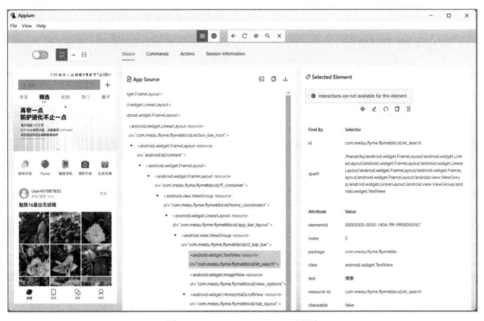

图 9-1　Appium Inspector 界面

9.2.2 Appium 的使用

启动 Appium 服务，默认监听本地 4723 端口。

```
> appium server --address '127.0.0.1' -p 4723
...
[Appium] Welcome to Appium v2.2.2
[Appium] Attempting to load driver espresso...
[Appium] Requiring driver at
C:\Users\fnngj\.appium\node_modules\appium-espresso-driver
[Appium] Attempting to load driver uiautomator2...
[Appium] Requiring driver at
C:\Users\fnngj\.appium\node_modules\appium-uiautomator2-driver
[Appium] Attempting to load driver xcuitest...
[Appium] Requiring driver at
C:\Users\fnngj\.appium\node_modules\appium-xcuitest-driver
[Appium] Appium REST http interface listener started on http://0.0.0.0:4723
[Appium] You can provide the following URLs in your client code to connect to this
server:
[Appium]        http://10.2.***.4:4723/
[Appium]        http://192.168.***.92:4723/
[Appium]        http://127.0.0.1:4723/ (only accessible from the same host)
[Appium]        http://172.22.***.1:4723/
[Appium] Available drivers:
[Appium]   - espresso@2.29.3 (automationName 'Espresso')
[Appium]   - uiautomator2@2.12.6 (automationName 'UiAutomator2')
[Appium]   - xcuitest@4.18.1 (automationName 'XCUITest')
[Appium] Available plugins:
[Appium]   - gestures@1.0.0-beta.4
[Appium]   - appium-reporter-plugin@1.0.0-beta.7
...
```

参数说明：

- --address：指定 IP 地址。
- -p：指定端口。

接下来编写 App 自动化测试用例。

```python
# appium_android_demo.py
from time import sleep
from appium import webdriver
from appium.options.android import UiAutomator2Options
from appium.webdriver.common.appiumby import AppiumBy


# 定义elems字典
elems = {
    "search_icon": "com.meizu.flyme.flymebbs:id/view_search",
    "search_input": "com.meizu.flyme.flymebbs:id/et_search",
    "result_title":
'//*[@resource-id="com.meizu.flyme.flymebbs:id/tv_post_title"]'
}
```

```python
# 配置Android App
capabilities = {
    "automationName": "UiAutomator2",
    "platformName": "Android",
    "appPackage": "com.meizu.flyme.flymebbs",
    "appActivity": "com.meizu.myplus.ui.splash.SplashActivity",
    "noReset": True,
}

appium_server_url = "http://127.0.0.1:4723"

options = UiAutomator2Options().load_capabilities(capabilities)
driver = webdriver.Remote(command_executor=appium_server_url, options=options)
driver.implicitly_wait(10)

# 搜索Flyme关键字
driver.find_element(AppiumBy.ID, elems.get("search_icon")).click()
driver.find_element(AppiumBy.ID, elems.get("search_input")).send_keys("flyme")
driver.execute_script('mobile: performEditorAction', {'action': 'search'})

# 打印结果列表
result_list = driver.find_elements(AppiumBy.XPATH, elems.get("result_title"))
for title in result_list:
    print(title.text)

# 关闭App
driver.quit()
```

代码说明

由于 Appium 的元素定位较长，所以这里通过定义 elems 字典来单独管理元素定位。

capabilities 定义了 App 的启动配置，其中，automationName 用于指定使用的 App 自动化测试库，platformName 用于指定手机系统，appPackage 和 appActivity 用于指定 App 的包名和主 Activity 名，noReset 表示在启动时不重置 App。

调用 UiAutomator2Options 类的 load_capabilities()方法，并传入 capabilities 配置字典，以准备自动化测试设置。随后，利用 webdriver.Remote()方法定义 WebDriver 实例。其中 command_executor 参数指定了 Appium 服务器的地址及端口，而 options 参数则包含了通过 UiAutomator2Options 类加载的启动配置信息。

接下来，基于 driver 对象操作元素，包括元素的点击和文本输入。execute_script()方法可用于模拟手机软键盘上的"搜索"按钮。最后，打印结果列表，退出驱动。

需要注意的是，随着时间的推移，本书中的例子可能会失效，因为 Appium 的 API 会变更，App 的元素定位及被测试功能也可能会变更。

9.3 Appium API 封装

本节主要介绍基于 Appium-Python-Client 库的 Appium API 封装技术。

Appium API 继承自 Selenium API。比如，Selenium 提供了 id、name、class、XPath 等定位，Appium 在此基础上扩展了 accessibility_id 和 ios_uiautomation 等针对 App 移动端设计的定位方式。再比如，Selenium 提供了 click()、send_keys()等操作方法，Appium 扩展了 top()、lang_press()、swipe()等操作方法。

Appium 中的有些操作写起来比较复杂，比如，屏幕滑动。

```
from selenium.webdriver.common.actions import interaction
from selenium.webdriver.common.action_chains import ActionChains
from selenium.webdriver.common.actions.action_builder import ActionBuilder

...

actions = ActionChains(driver)
# override as 'touch' pointer action
actions.w3c_actions = ActionBuilder(driver,
mouse=PointerInput(interaction.POINTER_TOUCH, "touch"))
actions.w3c_actions.pointer_action.move_to_location(start_x, start_y)
actions.w3c_actions.pointer_action.pointer_down()
actions.w3c_actions.pointer_action.pause(2)
actions.w3c_actions.pointer_action.move_to_location(end_x, end_y)
actions.w3c_actions.pointer_action.release()
actions.perform()
```

整个滑动过程是：触摸一个坐标位→按下→暂停→到另一个坐标位→释放→执行。由于过程较为烦琐，所以我们可以通过封装来简化这类操作。

📁 目录结构

```
09_chapter/
├──appium_lab
│   ├──__init__.py
│   ├──action.py
│   ├──find.py
│   ├──keyevent.py
│   └──switch.py
└──test_appium_lab.py
```

9.3.1 Switch 类

移动 App 可以根据其技术架构分为原生应用、Web 应用、Flutter 应用和混合应用。对于大型 App 而言，它们往往采用混合应用。比如，一个 App 可能最初使用原生 UI 进行开发，但某些页面为了可以灵活、实时地调整 UI 布局，会嵌入 WebView。随着 Flutter 技术的普及，为了提高开发效率并降低成本,很多新开发的页面会选择采用 Flutter 来实现。

原生 UI、Web UI 和 Flutter UI 在元素定位时是不同的,这就涉及 context 切换,即在不同的页面要切换到不同的模式,这样才能使用相应的定位方式来操作页面元素。

下面实现 Switch 类,以实现不同模式的 context 切换。

```python
# appium_lab/switch.py

class Switch:
    """
    基于 Appium 实现 Context 切换
    """

    def __init__(self, driver):
        self.driver = driver

    def context(self) -> str:
        """
        返回当前 context
        :return
        """
        current_context = self.driver.current_context
        all_context = self.driver.contexts
        print(f"current context: {current_context}.")
        print(f"all context: {all_context}.")
        return current_context

    def switch_to_app(self) -> None:
        """
        切换到原生 App
        """
        current_context = self.driver.current_context
        if current_context != "NATIVE_APP":
            print("Switch to native.")
            self.driver.switch_to.context('NATIVE_APP')

    def switch_to_web(self, context_name: str = None) -> None:
        """
        切换到 WebView
        :param context_name: WebView 上下文名称
        :return
        """
        print("Switch to webview.")
        if context_name is not None:
            self.driver.switch_to.context(context_name)
        else:
            all_context = self.driver.contexts
            for context in all_context:
                if "WEBVIEW" in context:
```

```python
                self.driver.switch_to.context(context)
                break
        else:
            raise NameError("No WebView found.")

def switch_to_flutter(self) -> None:
    """
    切换到 Flutter
    :return
    """
    current_context = self.driver.current_context
    if current_context != "FLUTTER":
        print("Switch to flutter.")
        self.driver.switch_to.context('FLUTTER')
```

📖 代码说明

创建 Switch 类，__init__()初始化方法用于接收 driver 驱动，current_context 语句用于获取当前上下文。Native 和 Flutter 可以根据固定的名字切换，WebView 需要通过 contexts 语句获取所有上下文，然后进一步判断是否包含"WEBVIEW"关键字。

使用示例

下面通过示例演示 Switch 类的使用。

```python
# test_appium_lab.py
from appium.webdriver import Remote
from appium.options.android import UiAutomator2Options
from appium_lab.switch import Switch

capabilities = {
    ...
}

appium_server_url = "http://127.0.0.1:4723"
options = UiAutomator2Options().load_capabilities(capabilities)
driver = Remote(command_executor=appium_server_url, options=options)
driver.implicitly_wait(10)

context = Switch(driver)
# 打印并返回当前上下文
context.context()
# 切换到 WebView
context.switch_to_web()
# 切换到 Native
context.switch_to_app()
# 切换到 Flutter
context.switch_to_flutter()
```

我们需要准备一款混合应用的 App，并设置 App 启动参数 capabilities。

⌛ 运行结果

```
> python test_appium_lab.py
current context: NATIVE_APP.
all context: ['NATIVE_APP', 'WEBVIEW_com.huawei.browser'].
Switch to webview....
Switch to native app....
Switch to flutter.
```

9.3.2　Action 类

Action 类中提供了基本的滑动和触摸操作。

```python
# appium_lab/action.py
from time import sleep
from selenium.webdriver.common.action_chains import ActionChains
from selenium.webdriver.common.actions.action_builder import ActionBuilder
from selenium.webdriver.common.actions.pointer_input import PointerInput
from selenium.webdriver.common.actions import interaction
from appium_lab.switch import Switch


class Action(Switch):
    """
    封装滑动和触摸操作
    """

    def __init__(self, driver):
        Switch.__init__(self, driver)
        self.switch_to_app()
        self._size = self.driver.get_window_size()
        self.width = self._size.get("width")
        self.height = self._size.get("height")

    def size(self) -> dict:
        """
        返回屏幕尺寸
        """
        print(f"screen resolution: {self._size}")
        return self._size

    def tap(self, x: int, y: int) -> None:
        """
        触摸坐标位
        :param x: x 坐标
        :param y: y 坐标
        :return:
```

```python
        """
        self.switch_to_app()
        print(f"top x={x},y={y}.")
        actions = ActionChains(self.driver)
        actions.w3c_actions = ActionBuilder(
            self.driver,
            mouse=PointerInput(interaction.POINTER_TOUCH, "touch"))
        actions.w3c_actions.pointer_action.move_to_location(x, y)
        actions.w3c_actions.pointer_action.pointer_down()
        actions.w3c_actions.pointer_action.pause(0.1)
        actions.w3c_actions.pointer_action.release()
        actions.perform()
        sleep(2)

    def swipe_up(self, times: int = 1, upper: bool = False) -> None:
        """
        向上滑动
        :param times: 滑动次数，默认值为1
        :param upper: 由于屏幕被键盘遮挡，所以可以选择滑动屏幕的上半部分
        :return:
        """
        self.switch_to_app()
        print(f"swipe up {times} times")
        x_start = int(self.width / 2)
        x_end = int(self.width / 2)

        if upper is True:
            self.height = (self.height / 2)

        y_start = int((self.height / 3) * 2)
        y_end = int((self.height / 3) * 1)

        for _ in range(times):
            actions = ActionChains(self.driver)
            actions.w3c_actions = ActionBuilder(
                self.driver,
                mouse=PointerInput(interaction.POINTER_TOUCH, "touch"))
            actions.w3c_actions.pointer_action.move_to_location(x_start, y_start)
            actions.w3c_actions.pointer_action.pointer_down()
            actions.w3c_actions.pointer_action.move_to_location(x_end, y_end)
            actions.w3c_actions.pointer_action.release()
            actions.perform()
            sleep(1)

    def swipe_down(self, times: int = 1, upper: bool = False) -> None:
        """
        向下滑动
        :param times: 滑动次数，默认值为1
        :param upper: 由于屏幕被键盘遮挡，所以可以选择仅滑动屏幕的上半部分
```

```
        :return:
        """
        self.switch_to_app()
        print(f"swipe down {times} times")
        x_start = int(self.width / 2)
        x_end = int(self.width / 2)

        if upper is True:
            self.height = (self.height / 2)

        y_start = int((self.height / 3) * 1)
        y_end = int((self.height / 3) * 2)

        for _ in range(times):
            actions = ActionChains(self.driver)
            actions.w3c_actions = ActionBuilder(
                self.driver,
                mouse=PointerInput(interaction.POINTER_TOUCH, "touch"))
            actions.w3c_actions.pointer_action.move_to_location(x_start, y_start)
            actions.w3c_actions.pointer_action.pointer_down()
            actions.w3c_actions.pointer_action.move_to_location(x_end, y_end)
            actions.w3c_actions.pointer_action.release()
            actions.perform()
            sleep(1)
```

代码说明

Action 类继承自 Switch 类，因为相关操作基于原生模式，所以需要调用 Switch 类封装的 switch_to_app()方法。

get_window_size()方法用于获取当前屏幕的 x 和 y 坐标，因为触摸、滑动等相关操作是基于坐标完成的。

top()方法用于接收 x 和 y 坐标，然后进行触摸操作。

swipe_up()方法和 swipe_down()方法是根据屏幕的大小进行计算的。x 轴居中，向上滑动 y 轴，从屏幕高度的 2/3 位置处滑动到 1/3 位置处，向下滑动则相反。通过 times 参数可以控制滑动次数。当屏幕键盘处于弹出状态时，它会占用屏幕底部大约 1/3 的空间，此时可以通过 upper 参数来控制仅滑动屏幕的上半部分。

使用示例

下面通过示例演示 Action 类的使用。

```
# test_appium_lab.py
from appium.webdriver import Remote
from appium.options.android import UiAutomator2Options
from appium_lab.action import Action
```

```
capabilities = {
    ...
}

appium_server_url = "http://127.0.0.1:4723"
options = UiAutomator2Options().load_capabilities(capabilities)
driver = Remote(command_executor=appium_server_url, options=options)
driver.implicitly_wait(10)

action = Action(driver)
# 屏幕尺寸
action.size()
# 向上滑动
action.swipe_up(times=3)
# 向下滑动
action.swipe_down(times=1)
# 触摸坐标位
action.tap(x=100, y=1333)
```

⏳ 运行结果

```
> python test_appium_lab.py
screen resolution: {'width': 1200, 'height': 2486}
swipe up 3 times
swipe down 1 times
top x=100,y=1333.
```

9.3.3 FindByText 类

一个元素可以没有 id 和 name，但是，但很可能有显示的文本，下面实现一组基于文本查找元素的方法。

某元素的元素属性列表如图 9-2 所示，我们可以基于 resource-id 来定位元素。

在实际编写 App 自动化测试时，经常会遇到没有定义 resource-id 的情况，特别是在 iOS 平台上，App 可能没有 id 属性，但一定会有 class 属性，因为 class 属性表示控件的类型。text 或 content-desc 用于显示控件名（iOS 平台上的 App 一般用 name 显示控件名），控件名一般是会有的。因此，基于 class 和 text，大概率可以定位到元素。

Attribute	Value
elementId	00000000-0000-1404-ffff-ffff00000067
index	5
package	com.meizu.flyme.flymebbs
class	android.widget.TextView
text	综合讨论
resource-id	com.meizu.flyme.flymebbs:id/tv_ad_res1
checkable	false
checked	false
clickable	true
enabled	true
focusable	true
focused	false
long-clickable	false
password	false
scrollable	false
selected	false
bounds	[53,1140][213,1214]

图 9-2　元素属性列表

下面实现 Find ByText 类，用于封装基于文本查找元素的功能。

```python
# appium_lab/find.py
from time import sleep
from appium.webdriver.common.appiumby import AppiumBy
from appium_lab.switch import Switch


class FindByText(Switch):
    """
    基于文本查找元素
    """

    def __find(self, class_name: str, attribute: str, text: str):
        """
        查找元素
        :param class_name: class 名字
        :param attribute: 属性
        :param text: 文本
        """
        elems = self.driver.find_elements(AppiumBy.CLASS_NAME, class_name)
        for _ in range(3):
            if len(elems) > 0:
                break
            sleep(1)
```

```python
        for elem in elems:
            if elem.get_attribute(attribute) is None:
                continue
            attribute_text = elem.get_attribute(attribute)
            if text in attribute_text:
                print(f'find -> {attribute_text}')
                return elem
        return None

    def find_text_view(self, text: str) -> None:
        """
        Android：基于 TextView 查找文本
        :param text: 文本名
        """
        self.switch_to_app()
        for _ in range(3):
            elem = self.__find(class_name="android.widget.TextView",
                               attribute="text", text=text)
            if elem is not None:
                break
            sleep(1)
        else:
            raise ValueError(f"Unable to find -> {text}")

        return elem

    def find_text_field(self, text: str) -> None:
        """
        iOS：基于 XCUIElementTypeTextField 查找文本
        :param text: 文本名
        """
        self.switch_to_app()
        for _ in range(3):
            elem = self.__find(class_name="XCUIElementTypeTextField",
                               attribute="name", text=text)
            if elem is not None:
                break
            sleep(1)
        else:
            raise ValueError(f"Unable to find -> {text}")
```

📖 **代码说明**

FindByText 类同样继承自 Switch 类，因为相关操作都基于原生模式，所以需要调用 Switch 类封装的 switch_to_app()方法。

self.__find()为类内部方法，它首先根据元素的 class 属性查找页面上的所有相关元素，随后在这些元素中，利用它们的文本属性（在 Android 中为 text 或 content-desc，在 iOS 中通常为 name）

来进一步筛选。具体操作是遍历这些元素，并检查它们的文本属性显示的名称是否包含指定的 text 字符串。一旦找到匹配项，便将其返回。

find_text_view()方法是针对 Android 平台的封装，它基于 android.widget.TextView 类来操作文本控件。

而 find_text_field()方法则是专为 iOS 平台设计的，它基于 XCUIElementTypeTextField 类来实现对文本输入框控件的操作。

在 Android 和 iOS 平台中带文本的组件列表如表 9-1 所示。

表 9-1 带文本的组件列表

平台	类名	文本属性
Android	android.view.View	Text或content-desc
Android	android.widget.EditText	text
Android	android.widget.Button	Text或content-desc
Android	android.widget.TextView	text
Android	android.widget.ImageView	content-desc
Android	android.widget.CheckBox	text
iOS	XCUIElementTypeStaticText	name
iOS	XCUIElementTypeOther	name
iOS	XCUIElementTypeTextField	name
iOS	XCUIElementTypeImage	name
iOS	XCUIElementTypeButton	name

使用示例

下面通过示例演示 Find ByText 类的使用。

```
# test_appium_lab.py
from appium.webdriver import Remote
from appium.options.android import UiAutomator2Options
from appium_lab.find import FindByText

capabilities = {
    ...
}

appium_server_url = "http://127.0.0.1:4723"
options = UiAutomator2Options().load_capabilities(capabilities)
driver = Remote(command_executor=appium_server_url, options=options)
driver.implicitly_wait(10)
```

```python
# 使用文本定位
find = FindByText(driver)
find.find_text_view("综合讨论").click()
```

运行结果

```
> python test_appium_lab.py
find -> 综合讨论
```

9.3.4 KeyEvent 类

Appium 提供了两种输入字符编码的方法：一种是从 Selenium 继承而来的 send_keys()方法，另一种是调用屏幕键盘的 press_keycode()方法。

press_keycode()方法的输入非常不方便，比如，想要输入"HELLO"字符串，就需要通过 Android KeyEvent 文档查询每个字符对应的编码。

"HELLO"字符串分别对应：H - 36，E - 33，L - 40，O - 43。

随后，通过 press_keycode()方法分别输入编码。

```
driver.keyevent(36)
driver.keyevent(33)
driver.keyevent(40)
driver.keyevent(40)
driver.keyevent(43)
```

KeyEvent 类用于封装基于文本查找元素的功能。

```python
# appium_lab/keyevent.py

keycodes = {
    '0': 7, '1': 8, '2': 9, '3': 10, '4': 11, '5': 12, '6': 13, '7': 14, '8': 15,
    '9': 16, 'A': 29, 'B': 30, 'C': 31, 'D': 32, 'E': 33, 'F': 34, 'G': 35, 'H': 36,
    'I': 37, 'J': 38, 'K': 39, 'L': 40, 'M': 41, 'N': 42, 'O': 43, 'P': 44, 'Q': 45,
    'R': 46, 'S': 47, 'T': 48, 'U': 49, 'V': 50, 'W': 51, 'X': 52, 'Y': 53, 'Z': 54,
    ' ': 62, '*': 17, '#': 18, ',': 55, '`': 68, '-': 69, '[': 71, ']': 72, '\\': 73,
    ';': 74, '/': 76, '@': 77, '=': 161, '.': 158, '+': 157,
    'NUM_LOCK': 143, 'CAPS_LOCK': 115, 'HOME': 4, 'BACK': 3, 'ENTER': 66,
}


class KeyEvent:
    """
    KeyEvent:
    https://developer.andr***.com/reference/android/view/KeyEvent
    """

    def __init__(self, driver):
        self.driver = driver
```

```python
def key_text(self, text: str = ""):
    """
    通过键盘输入文本
    :param text: 输入字符串

    Usage:
        key_text("Hello")
    """
    if text == "":
        return

    print(f'input "{text}"')
    for string in text:
        keycode = keycodes.get(string.upper(), 0)
        if keycode == 0:
            raise KeyError(f"The '{string}' character is not supported")
        if string.isupper():
            self.driver.press_keycode(keycode, 64, 59)
        else:
            self.driver.keyevent(keycode)

def press_key(self, key: str):
    """
    输入某个键名
    :param key: 键名
    press_key("HOME")
    """
    print(f'press key "{key}"')
    keycode = keycodes.get(key.upper(), 0)
    if keycode == 0:
        raise KeyError(f"The '{key}' character is not supported")
    self.driver.press_keycode(keycode)
```

📖 代码说明

keycodes 字典用于保存字符与编码之间的对应关系。

KeyEvent 类中实现了 key_text() 和 press_key() 方法。

key_text() 方法可模拟键盘输入字符串。它会循环遍历字符串中的每个字符并逐一输入。当需要输入大写字母时，它首先利用 upper() 方法将整个字符串转换为大写形式，接着通过 keycodes 字典查来查找每个大写字母对应的键盘编码，并调用 press_keycode() 方法输入大写字母。相反，如果字母为小写，则直接通过 keycodes 字典查找每个小写字母对应的键盘编码，并使用 keyevent() 方法输入小写字母。

press_key() 方法用于输入手机按键，比如，Home 键和 Back 键。

🖱 使用示例

下面通过示例演示 KeyEvent 类的使用。

```python
# test_appium_lab.py
from time import sleep
from appium.webdriver import Remote
from appium_lab.keyevent import KeyEvent
from appium.webdriver.common.appiumby import AppiumBy
from appium.options.android import UiAutomator2Options

capabilities = {
    ...
}

appium_server_url = "http://127.0.0.1:4723"
options = UiAutomator2Options().load_capabilities(capabilities)
driver = Remote(command_executor=appium_server_url, options=options)
driver.implicitly_wait(10)

# 键盘输入
key = KeyEvent(driver)
key.key_text("Flyme10")
sleep(1)
key.press_key("ENTER")
```

⌛ 运行结果

```
> python test_appium_lab.py
input "Flyme10"
press key "ENTER"
```

9.4　Appium 图像与文字识别

在进行 App 自动化测试过程中，由于页面的复杂性，我们往往需要根据页面的技术实现（Native、MWeb、Flutter）使用不同的工具。而频繁切换定位工具给自动化测试的稳定性带来了不小的挑战。图像与文字识别技术不依赖于页面本身是用什么技术实现的，只要能识别图片上的元素或文字并生成坐标即可定位，这将是未来探索自动化测试技术的重要方向之一。Appium 提供了一些插件，这些插件可以利用图像或文字识别来定位元素。

📁 目录结构

```
09_chapter/
├──image
│   └──phone.jpg
├──extension
│   ├──__init__.py
```

```
    │    └──ocr_extension.py
    ├──test_appium_image.py
    └──test_appium_orc.py
```

9.4.1 images 插件

在 Appium 中，通过使用 images 插件的"-image"定位功能，可以指定想要定位的元素的图片文件。如果 Appium 可以找到与图片匹配的屏幕区域，则它会将该区域的相关信息包装为标准的 WebElement，并将其发送给 Appium 客户端。

1. 使用插件

在使用"-image"定位之前，应安装 Appium 插件。

```
> appium plugin install images
```

查看已安装的 Appium 插件。

```
> appium plugin list --installed
✔ Listing installed plugins
- images@2.1.8 [installed (npm)]
```

启动 Appium Server，并指定使用 images 插件。

```
> appium server --address '127.0.0.1' -p 4723 --use-plugins=images
```

App 首页如图 9-3 所示。

图 9-3 App 首页

首先，对想要定位的元素进行截图，并保存为名为"phone.jpg"的图片文件，如图 9-4 所示。

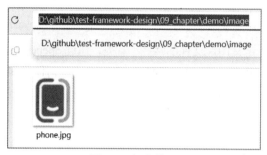

图 9-4　保存截图

然后，编写 Appium 自动化测试脚本。

```
# test_appium_image.py
import os
import base64
from appium import webdriver
from appium.options.android import UiAutomator2Options
from appium.webdriver.common.Appiumby import AppiumBy


capabilities = {
    "automationName": "UiAutomator2",
    "platformName": "Android",
    "appPackage": "com.meizu.flyme.flymebbs",
    "appActivity": "com.meizu.myplus.ui.splash.SplashActivity",
    "noReset": True,
}

Appium_server_url = "http://127.0.0.1:4723"
options = UiAutomator2Options().load_capabilities(capabilities)
driver = webdriver.Remote(command_executor=Appium_server_url, options=options)

driver.update_settings({"fixImageTemplatescale": True})
driver.implicitly_wait(10)

current_dir = os.path.dirname(os.path.abspath(__file__))
image_path = os.path.join(current_dir, "image", "phone.jpg")

with open(image_path, 'rb') as png_file:
    b64_data = base64.b64encode(png_file.read()).decode('UTF-8')

driver.find_element(AppiumBy.IMAGE, b64_data).click()
```

📖 **代码说明**

启动 App 的过程与前面的示例相同，下面重点介绍图片定位部分。

update_settings()方法用于设置当前会话，将 fixImageTemplatescale 设置为 True。

image_path 变量用于定义图片文件的路径。首先，通过 open()方法打开图片文件，用 read()方法读取图片文件内容。其次，使用 base64.b64encode()方法对图片文件进行编码。最后，使用"AppiumBy.IMAGE"识别图片在整个页面上的坐标位置，并对坐标位置进行 click()操作。

2. 图像元素操作

仅通过图片就能定位元素的操作如下。

- Click。
- isDisplayed。
- getSize。
- getLocation。
- getLocationInView。
- getElementRect。
- getAttribute。
  - ◎ visual。
  - ◎ score。

这些操作可以基于"图片元素"进行定位，是因为它们仅涉及屏幕位置。而其他操作（比如 sendKeys）不支持"图片元素"，是因为根据提供的定位图片，Appium 只知道是否有一个屏幕区域与其视觉匹配，但无法将这些信息转化为特定驱动程序的 UI 元素对象，所以无法实现元素坐标的输入。

以上操作在 Appium-Python-Client 库中的 API 如下。

```python
# appium_image_demo.py

...
driver.update_settings({"getMatchedImageResult": True})

...

element = driver.find_element(AppiumBy.IMAGE, b64_data)

# 点击
element.click()

# 是否显示
print(element.is_displayed())
```

```
# 获取尺寸
print(element.size)

# 元素在可渲染画布中的位置
print(element.location)

# 获取元素相对于视图的位置
print(element.location_in_view)

# 获取坐标和尺寸
print(element.rect)

# 将返回的匹配的图像作为base64数据，需要设置getMatchedImageResult为True
print(element.get_attribute("visual"))

# 自Appium 1.18.0起，返回的相似度分数为浮点数，范围在[0.0, 1.0]区间
print(element.get_attribute("score"))
```

3. 相关设置

通过图像查找元素依赖于图像分析软件与Appium的截图功能，以及提供的图片。Appium提供了一些参数，可以调节此功能，在某些情况下可以加快匹配速度或使其更准确，如表9-2所示。

表9-2 Appium提供的参数

参数名称	说明	可能值	默认值
imageMatchThreshold	OpenCV的匹配阈值。如果低于这个阈值，则认为查找失败。通常来说，可能值在0（意味着不使用阈值）到1（意味着参考图片必须是完全像素对像素的匹配）之间	0到1之间	0.4
fixImageFindScreenshotDims	如果检索到的屏幕截图与屏幕尺寸不匹配，则这个设置决定了Appium会调整屏幕截图的尺寸来匹配，以确保在正确的坐标上找到匹配的元素	True 或 False	True
fixImageTemplateSize	如果一个参考图片或模板的尺寸大于要匹配的图片，则OpenCV将不允许匹配该图片。可能发生的情况是，我们发送的参考图片的尺寸比Appium检索的截图大。在这种情况下，匹配会自动失败。如果把这个参数设置为True，则Appium会自动调整模板的大小，以确保它比截图的尺寸小	True 或 False	False

续表

参数名称	说明	可能值	默认值
fixImageTemplateScale	Appium在与OpenCV匹配之前，会调整基础图片的大小以适应其窗口大小。如果把这个参数设置为True，则Appium会把我们发送的参考图片按同样的比例进行缩放，以适应窗口大小。比如，截图是750像素×1334像素的基础图片，当窗口大小为375像素×667像素时，Appium会将基础图像图片重新缩小为窗口大小，缩放比例为0.5。参考图片一般基于屏幕截图尺寸，从来没有图像与窗口尺寸的比例。这个设置允许Appium用0.5来缩放参考图片	True 或 False	False
defaultImageTemplatescale	Appium默认不调整模板图片的大小（值为1.0）。虽然，存储缩放的模板图片可能有助于节省存储空间。比如，可以用270像素×32像素的模板图片来表示1080像素×126像素的区域（defaultImageTemplatescale的值被设置为4.0）	比如，0.5、1.0	1.0
checkForImageElementStaleness	可能发生的情况是，在匹配了一个图像元素之后和在选择点击它之前，这个元素已经不存在了。Appium判断这一点的唯一方法就是在此之前尝试重新匹配模板。如果重新匹配失败，则会得到一个StaleElementException。如果将这个选项设置为False，则可以跳过此项检查，这可能会加快检查速度，但也有可能会遇到问题，且不出现异常	True 或 False	True
autoUpdateImageElementPosition	在匹配了一个图像元素之后和点击它之前，这个元素的位置发生了变化。如果Appium在重新匹配时确定位置改变了，那么它可以自动调整位置	True 或 False	False
imageElementTapStrategy	为了点击找到的图片元素，Appium必须使用一个触摸动作策略。可用的策略是W3C Actions API或旧的MJSONWP TouchActions API。除非使用的驱动因某些原因不支持W3C Actions API，否则请坚持使用默认值	W3C Actions 或 MJSONWPTouchActions	W3C Actions
getMatchedImageResult	Appium不存储匹配的结果。当将结果存储在内存中时，可能会有助于调试是否有哪个区域被find by image匹配。Appium会将属性API中的图像作为visual返回	True 或 False	False

参考前面的完整示例，用法如下。

```
#appium_image_demo.py

...
driver.update_settings({"getMatchedImageResult": True})
driver.update_settings({"fixImageTemplatescale": True})
```

9.4.2　Appium OCR 插件

这是一个基于 Tesseract 的 Appium OCR 插件。它依赖于 Tesseract.js 进行 OCR 处理，其特点如下。

- 新的 OCR 端点：调用新的 Appium 服务端点对当前截图进行 OCR，并返回匹配的文本和元数据。
- OCR context：当切换到 OCR 上下文时，系统会更新页面源信息，使其包含屏幕上识别出的文本对象的 XML 表示，从而与这些文本元素进行交互。
- 基于 OCR 文本查找元素：在 OCR 上下文中，XPath 将根据屏幕内容的 OCR 生成的 XML 表示来定位"元素"。随后，与这些识别出的元素的交互（如 click、getText）均基于它们在屏幕上的绝对位置进行，以实现精准的最小化交互。

1. 使用插件

安装 Appium OCR 插件。

```
> appium plugin install images--source=npm appium-ocr-plugin
```

查看已安装的插件。

```
> appium plugin list --installed
✔ Listing installed plugins
- ocr@0.2.0 [installed (npm)]
```

在启动 Appium Server 时指定使用 Appium OCR 插件。

```
> appium server --address '127.0.0.1' -p 4723 --use-plugins=ocr
```

创建 OCR 扩展脚本。

```python
# extension/ocr_extension.py
from appium.webdriver.webdriver import ExtensionBase

# define an extension class
class OCRCommand(ExtensionBase):
    def method_name(self):
        return 'ocr_command'

    def ocr_command(self, argument):
        return self.execute(argument)['value']

    def add_command(self):
        add = ('post', '/session/$sessionId/appium/ocr')
        return add
```

📖 代码说明

创建 OCRCommand 类，该类继承自 ExtensionBase 类。ExtensionBase 类作为基础，用于将

自定义扩展命令集成到驱动程序中。在 OCRCommand 类中，method_name()方法用于设定命令的名称，ocr_command()方法用于实现具体的命令逻辑，add_command()方法则用于向驱动程序注册此自定义命令。

如果想进一步了解 Appium 扩展的开发，则可以查看 ExtensionBase 类的内部注释，这里不再展开介绍。

编写 App 自动化测试脚本。

```
# test_appium_ocr.py
from appium import webdriver
from appium.options.android import UiAutomator2Options
from appium.webdriver.common.appiumby import AppiumBy
from extension.ocr_extension import OCRCommand

capabilities = {
    "automationName": "UiAutomator2",
    "platformName": "Android",
    "appPackage": "com.meizu.flyme.flymebbs",
    "appActivity": "com.meizu.myplus.ui.splash.SplashActivity",
    "noReset": True,
}

appium_server_url = "http://127.0.0.1:4723"
options = UiAutomator2Options().load_capabilities(capabilities)
driver = webdriver.Remote(command_executor=appium_server_url, options=options,
extensions=[OCRCommand])
driver.implicitly_wait(10)

ocr = driver.ocr_command({})
print(ocr)
```

📖 代码说明

首先导入 OCRCommand 类，当通过 Remote 类创建驱动时，可以通过 extensions 参数传入 OCRCommand 类。

调用 ocr_command()方法，即 OCRCommand 类实现的方法，对当前屏幕进行截图，并通过 OCR 技术识别图片中的文字。

2. OCR 识别对象

根据前文的代码，打印 ocr 变量可以得到一个 JSON 结构体。

```
{
  "words": [
    {
      "text": "mEngine", "confidence": 88.47775268554688,
      "bbox": {"x0": 86, "y0": 509, "x1": 308, "y1": 560}
```

```
      },
      {
        "text": "Flyme", "confidence": 91.3454818725586,
         "bbox": {"x0": 316, "y0": 1132, "x1": 420, "y1": 1172}
      },
      {
        "text": "A9", "confidence": 34.86248779296875,
        "bbox": {"x0": 1017, "y0": 2565, "x1": 1078, "y1": 2595}
      }
    ],
    "lines": [
      {
        "text": "mEngine BY Ni0vEh 1 Bl\n\n", "confidence": 21.003677368164062,
       "bbox": {"x0": 86, "y0": 500, "x1": 674, "y1": 560}
      },
      {
        "text": "Flyme\n\n", "confidence": 91.3454818725586,
       "bbox": {"x0": 316, "y0": 1132, "x1": 420, "y1": 1172}
      },
      {
        "text": "A9\n", "confidence": 34.86248779296875,
        "bbox": {"x0": 1017, "y0": 2565, "x1": 1078, "y1": 2595}
      }
    ],
    "blocks": [
      {
        "text": "mEngine BY Ni0vEh 1 Bl\n\n", "confidence": 21.003677368164062,
         "bbox": {"x0": 86, "y0": 500, "x1": 674, "y1": 560}
      },
      {
        "text": "Flyme\n\n", "confidence": 91.3454818725586,
        "bbox": {"x0": 316, "y0": 1132, "x1": 420, "y1": 1172}
      },
      {
        "text": "A9\n", "confidence": 34.86248779296875,
        "bbox": {"x0": 1017, "y0": 2565, "x1": 1078, "y1": 2595}
      }
    ]
}
```

说明如下。

- words：Tesseract 识别的单个单词的列表。
- lines：Tesseract 识别的文本行的列表。
- blocks：Tesseract 识别的连续文本块的列表。

每项都引用了一个 OCR 对象，它们本身包含三个数据。

- text：识别的文本。
- confidence：Tesseract 在处理给定文本后，会为 OCR 识别结果分配一个置信度值，该值范围在 0 到 100 之间。

- bbox：文本的边界框。边界框用于标识发现文本的区域，通过四个坐标值（$x0$、$y0$、$x1$、$y1$）来定义。其中，$x0$ 和 $y0$ 分别表示文本区域左上角的 x 坐标和 y 坐标，而 $x1$ 和 $y1$ 则分别表示文本区域右下角的 x 坐标和 y 坐标。$x0$ 到 $x1$ 定义了文本的水平宽度，$y0$ 和 $y1$ 定义了文本的垂直高度。

3. 查找与操作元素

当通过 OCR 技术识别出来了一些文本，并且知道这些文本的置信度值和坐标后，接下来就可以使用文本定位元素了，共分为两步。

第 1 步，切换 OCR 上下文。

第 2 步，通过 XPath 定位元素，并对其进行操作。

以某 App 首页为例，根据前面 OCR 的识别结果，识别到"Flyme"文本，如图 9-5 所示。

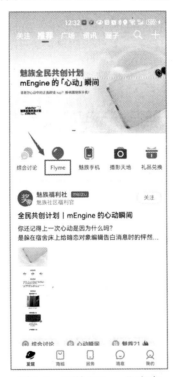

图 9-5 识别"Flyme"文本

示例代码如下。

```
# appium_ocr_demo.py
from appium import webdriver
```

```
from appium.options.android import UiAutomator2Options
from appium.webdriver.common.appiumby import AppiumBy
from extension.ocr_extension import OCRCommand

capabilities = {
    ...
}

appium_server_url = "http://127.0.0.1:4723"
options = UiAutomator2Options().load_capabilities(capabilities)
driver = webdriver.Remote(command_executor=appium_server_url, options=options,
extensions=[OCRCommand])
driver.implicitly_wait(10)

ocr = driver.ocr_command({})
print(ocr)

driver.switch_to.context('OCR')
driver.find_element(AppiumBy.XPATH, '//words/item[text() = "Flyme"]').click()
```

📖 代码说明

使用 switch_to.context() 方法切换到 OCR 上下文。之后，使用 XPath 定位元素，其中，words 表示从单词列表中选取文本，item 表示某一项，text()="Flyme"表示文本为"Flyme"关键字。

通过 OCR 定位到的元素并非标准的 UI 元素（即 iOS UI 的 XCUIElementTypeText 或 Android UI 的 android.widget.TextView）。相反，它们只是一种"虚拟"元素，仅支持以下方法。

- Click Element：在选定图片的边界框的中心点执行点击操作。
- Is Element Displayed：始终返回 True，因为不显示元素，所以它不适合 OCR。
- Get Element Size：返回元素的尺寸。
- Get Element Location：返回元素的坐标。
- Get Element Rect：返回元素的坐标和尺寸。
- Get Element Text：返回 OCR 识别的文本（该文本与页面源代码中的文本一致）。
- Get Element Attribute：只能检索一个属性（置信度），返回置信度值。

以上操作在 Appium-Python-Client 库中的 API 如下。

```
# test_appium_ocr.py

...
driver.switch_to.context('OCR')
element = driver.find_element(AppiumBy.XPATH, '//words/item[text() = "Flyme"]')
# 点击
element.click()

# 是否显示
```

```python
print(element.is_displayed())

# 尺寸
print(element.size)

# 坐标
print(element.location)

# 坐标和尺寸
print(element.rect)

# 文本
print(element.text)

# 属性（仅支持 confidence）
print(element.get_attribute("confidence"))
```

第 10 章 HTTP 接口自动化测试设计

近几年，随着微服务架构在各个软件公司的落地，软件服务的数量成指数级增加，而不同服务之间都需要通过接口通信，无形中软件的接口就变多了，整个系统的调用链也变得越发复杂，因此接口测试变得尤为重要。由于接口测试本身具有相对简单、稳定性好等特点，更加容易实现自动化测试，所以许多公司的测试团队会优先开展接口自动化测试。在主流的 Web 项目中，以 HTTP、RPC 接口为主，其中又以 HTTP 接口应用最为广泛。

10.1 HTTP 客户端库

在 Python 中有许多简单好用的 HTTP 测试库，比如，requests、HTTPX 和 aiohttp。接下来，我们逐一进行介绍。

10.1.1 requests

requests 是一个非常流行的 HTTP 客户端库。它是 Python 领域内的明星项目，凭借简单而强大的 API 被广泛应用于诸多领域，并且为其他 HTTP 客户端库的设计提供了模板。

用 pip 命令安装 requests。

```
> pip install requests
```

🔗 使用示例

下面通过 GET 方法实现一个接口调用，并且需要 auth 认证。

```
# requests_demo.py
import requests

r = requests.get('https://api.git***.com/user', auth=('user', 'pass'))
```

```
print(f"返回状态码：{r.status_code}")
print(f"返回数据：{r.json()}")
```

10.1.2　HTTPX

HTTPX 是 Python 的下一代 HTTP 客户端库，它包括一个集成的命令行客户端，支持 HTTP/1.1 和 HTTP/2，并同时提供同步和异步 API。

用 pip 命令安装 HTTPX。

```
> pip install httpx
```

🔸 使用示例

HTTPX 的用法和 requests 类似，下面通过 GET 方法请求一个页面。

```python
# httpx_sync_demo.py
import httpx

r = httpx.get('https://www.example.org/')
print(f"返回状态码：{r.status_code}")
print(f"返回数据：{r.text}")
```

此外，HTTPX 支持异步调用。

```python
# httpx_async_demo.py
import asyncio
import httpx

async def main():
    async with httpx.AsyncClient() as client:
        r = await client.get('https://www.example.com/')
        print(f"状态码：{r.status_code}")
        print(f"返回数据：{r.text}")

asyncio.run(main())
```

10.1.3　aiohttp

aiohttp 是一款支持异步 HTTP 客户端和服务端的框架，其特点如下。

- 支持 HTTP 的客户端和服务端。
- 支持客户端和服务端的 WebSocket，避免回调"地狱"。
- 提供具有中间件和可插拔路由的 Web 服务器。

用 pip 命令安装 aiohttp。

```
> pip install aiohttp
```

🖱 使用示例

下面基于 aiohttp 实现一个异步 Web 服务，使其包含 HTTP 服务和 WebSocket 服务。

```python
# aiohttp_server_demo.py
import aiohttp
from aiohttp import web

# HTTP 服务
async def http_handle(request):
    name = request.match_info.get('name', "Anonymous")
    text = "Hello, " + name
    return web.Response(text=text)


# WebSocket 服务
async def websocket_handler(request):

    ws = web.WebSocketResponse()
    await ws.prepare(request)

    async for msg in ws:
        if msg.type == aiohttp.WSMsgType.TEXT:
            if msg.data == 'close':
                await ws.close()
            else:
                await ws.send_str(f'Hello, {msg.data}')
        elif msg.type == aiohttp.WSMsgType.ERROR:
            print(f'ws connection closed with exception {ws.exception()}')

    print('websocket connection closed')

    return ws


app = web.Application()
app.add_routes([
    web.get('/hello/', http_handle),
    web.get('/hello/{name}', http_handle),
    web.get('/echo', websocket_handler)
])

if __name__ == '__main__':
    web.run_app(app, host="127.0.0.1", port=8080)
```

📖 代码说明

通过 web.Application 类创建一个 Web 应用，add_routes()方法用于添加 Web 路径。其中，/hello/路径指向 http_handle，/echo 路径指向 websocket_handler。run_app()方法用于启动 App， host 用于指定 IP 地址，默认为本机 IP 地址。port 用于指定端口，默认端口为 8080。

http_handle() 方法用于处理 HTTP 请求，request 参数用于接收请求数据，request.match_info.get() 方法用于获取 URI 路径中的 name。如果获取不到默认的 Anonymous，则 web.Response() 方法将返回字符串。

websocket_handler() 方法用于处理 WebSocket 连接，它创建了一个 WebSocketResponse 对象，赋值给了 ws，并使用 prepare() 方法来准备这个对象。在代码中使用 for 循环来接收消息，type 为消息的类型，data 为消息本身，send_str() 方法用于向客户端发送消息，close() 方法用于关闭 WebSocket 连接。

首先，启动服务。

```
> python aiohttp_server_demo.py
======== Running on http://127.0.0.1:8080 ========
(Press CTRL+C to quit)
```

其次，基于 aiohttp 实现一个异步 HTTP 客户端请求。

```python
# aiohttp_http_client_demo.py
import aiohttp
import asyncio

async def main():
    async with aiohttp.ClientSession() as session:
        async with session.get('http://127.0.0.1:8080/hello/jack') as response:
            print("Status:", response.status)
            print("Content-type:", response.headers['content-type'])
            text = await response.text()
            print("Response text:", text)

asyncio.run(main())
```

📖 代码说明

首先通过 ClientSession() 构造函数创建客户端 Session，然后基于该 Session 发送 GET 请求。在请求后，可以使用 response.status 参数获取 HTTP 状态码，使用 response.headers 参数获取响应头，使用 response.text() 方法获取返回的 text 文本信息。

▣ 运行结果

```
> python aiohttp_http_client_demo.py
Status: 200
Content-type: text/plain; charset=utf-8
Response text: Hello, jack
```

下面基于 aiohttp 实现一个异步 WebSocket 客户端请求。

```python
# aiohttp_ws_client_demo.py
import aiohttp
```

```
import asyncio

async def main():
    async with aiohttp.ClientSession() as session:
        async with session.ws_connect('ws://127.0.0.1:8080/echo') as ws:
            # 发送tom
            await ws.send_str("tom")
            async for msg in ws:
                print(f"ws client, data: {msg.data}")
                break

            # 发送jerry
            await asyncio.sleep(1)
            await ws.send_str("jerry")
            async for msg in ws:
                print(f"ws client, data: {msg.data}")
                break

            # 发送close
            await asyncio.sleep(1)
            await ws.send_str("close")

asyncio.run(main())
```

📖 **代码说明**

首先通过 ClientSession() 构造函数创建客户端 Session，然后基于该 Session 调用 ws_connect() 方法以发送 WebSocket 连接。在连接建立后，使用 send_str() 方法发送请求，使用 for 循环读取 ws 对象中的内容，使用 msg 获取服务端返回的消息对象，data 为消息文本。

asyncio.sleep() 方法用于异步中的休眠。

根据 WebSocket 服务端的判断逻辑，当发送"close"字符串时，会断开 WebSocket 连接。

需要说明的是，一旦建立了 WebSocket 连接，不管是服务端还是客户端，都可以发送消息，由另一端响应消息。当然，也可以由任意一端主动断开连接。

⌛ **运行结果**

```
> python aiohttp_ws_client_demo.py
ws client response, data: Hello, tom
ws server response, data: Hello, jerry
```

10.2 HTTP 请求方法集成日志

在进行 HTTP 接口自动化测试过程中，我们非常关心接口的请求信息和返回的结果。每次都通过 print() 函数进行打印其实并不方便，或者说效率不高。因此，我们可以集成日志，默认通过 logger 输出这些信息。

📁 目录结构

```
10_chapter/
├──common
│   ├──__init__.py
│   └──request_v1.py
└──test_request_v1.py
```

下面以 requests 为基础，重新封装 GET、POST、PUT 和 DELETE 这四个方法。

```python
import json
import requests
from loguru import logger


def formatting(msg):
    """格式化 JSON 数据"""
    if isinstance(msg, dict):
        return json.dumps(msg, indent=2, ensure_ascii=False)
    return msg


def request(func):
    """请求装饰器"""

    def wrapper(*args, **kwargs):
        func_name = func.__name__
        logger.info('-------------- 请求 -----------------')
        try:
            url = list(args)[1]
        except IndexError:
            url = kwargs.get("url", "")

        logger.info(f"[method]: {func_name.upper()}     [URL]: {url} ")
        auth = kwargs.get("auth", "")
        headers = kwargs.get("headers", "")
        cookies = kwargs.get("cookies", "")
        params = kwargs.get("params", "")
        data = kwargs.get("data", "")
        json_ = kwargs.get("json", "")
        if auth != "":
            logger.debug(f"[auth]:\n {auth}")
        if headers != "":
            logger.debug(f"[headers]:\n {formatting(headers)}")
        if cookies != "":
            logger.debug(f"[cookies]:\n {formatting(cookies)}")
        if params != "":
```

```python
                logger.debug(f"[params]:\n {formatting(params)}")
            if data != "":
                logger.debug(f"[data]:\n {formatting(data)}")
            if json_ != "":
                logger.debug(f"[json]:\n {formatting(json_)}")

            # running function
            r = func(*args, **kwargs)

            status_code = r.status_code
            logger.info("-------------- 响应 ----------------")
            if status_code == 200 or status_code == 304:
                logger.info(f"successful with status {status_code}")
            else:
                logger.warning(f"unsuccessful with status {status_code}")
            resp_time = r.elapsed.total_seconds()
            try:
                resp = r.json()
                logger.debug(f"[type]: json     [time]: {resp_time}")
                logger.debug(f"[response]:\n {formatting(resp)}")
            except BaseException as msg:
                logger.debug("[warning]: failed to convert res to json, try to convert to text")
                logger.trace(f"[warning]: {msg}")
                r.encoding = 'utf-8'
                logger.debug(f"[type]: text     [time]: {resp_time}")
                logger.debug(f"[response]:\n {r.text}")

            return r

    return wrapper


class HttpRequest:
    """http request class"""

    @request
    def get(self, url, params=None, **kwargs):
        return requests.get(url, params=params, **kwargs)

    @request
    def post(self, url, data=None, json=None, **kwargs):
        return requests.post(url, data=data, json=json, **kwargs)

    @request
    def put(self, url, data=None, **kwargs):
        return requests.put(url, data=data, **kwargs)

    @request
    def delete(self, url, **kwargs):
        return requests.delete(url, **kwargs)
```

📖 代码说明

在 HttpRequests 类中重新实现 GET、POST、PUT 和 DELETE 这四个方法，分别调用 requests 提供的四个同名方法，参数保持一致，并使用 request 装饰器进行装饰。

在 request 装饰器中获得请求和响应的相关数据。

- 请求：url、headers、auth 和 cookies，以及请求参数 params、data 和 json 等。
- 响应：响应数据的类型、响应时间和响应数据等。

利用 Loguru 日志库打印日志。如果判断数据为 JSON 格式，则调用 formatting()方法进行格式化输出。

🔧 使用示例

```
from common.request_v1 import httpRequest

http_req = HttpRequest()
http_req.get('http://http***.org/get', params={'key': 'value'})
http_req.post('http://http***.org/post', data={'key': 'value'})
http_req.put('http://http***.org/put', data={'key': 'value'})
http_req.delete('http://http***.org/delete')
```

⏳ 运行结果

```
> python test_request_v1.py
2024-01-06 23:45:26.213 | INFO     | common.request_v1:wrapper:18 - --------- 请求 ------------
2024-01-06 23:45:26.214 | INFO     | common.request_v1:wrapper:24 - [method]: GET [URL]: http://http***.org/get
2024-01-06 23:45:26.214 | DEBUG    | common.request_v1:wrapper:38 - [params]:
{
  "key": "value"
}
2024-01-06 23:45:26.787 | INFO     | common.request_v1:wrapper:48 - --------- 响应 ------------
2024-01-06 23:45:26.787 | INFO     | common.request_v1:wrapper:50 - successful with status 200
2024-01-06 23:45:26.787 | DEBUG    | common.request_v1:wrapper:56 - [type]: json [time]: 0.570168
2024-01-06 23:45:26.788 | DEBUG    | common.request_v1:wrapper:57 - [response]:
{
  "args": {
    "key": "value"
  },
  "headers": {
    "Accept": "*/*",
    "Accept-Encoding": "gzip, deflate",
    "Host": "httpbin.org",
    "User-Agent": "python-requests/2.31.0",
```

```
      "X-Amzn-Trace-Id": "Root=1-65997597-323fad017631a4df069dc87c"
    },
    "origin": "203.176.241.236",
    "url": "http://http***.org/get?key=value"
}
2024-01-06 23:45:26.788 | INFO     | common.request_v1:wrapper:18 - --------请求
-------------
2024-01-06 23:45:26.789 | INFO     | common.request_v1:wrapper:24 - [method]: POST [URL]: http://http***.org/post
2024-01-06 23:45:26.789 | DEBUG    | common.request_v1:wrapper:40 - [data]:
{
  "key": "value"
}
2024-01-06 23:45:27.302 | INFO     | common.request_v1:wrapper:48 - -------- 响应
------------
2024-01-06 23:45:27.302 | INFO     | common.request_v1:wrapper:50 - successful with status 200
2024-01-06 23:45:27.302 | DEBUG    | common.request_v1:wrapper:56 - [type]: json [time]: 0.510918
2024-01-06 23:45:27.302 | DEBUG    | common.request_v1:wrapper:57 - [response]:
{
  "args": {},
  "data": "",
  "files": {},
  "form": {
    "key": "value"
  },
  "headers": {
    "Accept": "*/*",
    "Accept-Encoding": "gzip, deflate",
    "Content-Length": "9",
    "Content-Type": "application/x-www-form-urlencoded",
    "Host": "httpbin.org",
    "User-Agent": "python-requests/2.31.0",
    "X-Amzn-Trace-Id": "Root=1-65997598-0546d8e3500e8b5e6ff13f48"
  },
  "json": null,
  "origin": "203.176.241.236",
  "url": "http://http***.org/post"
}
2024-01-06 23:45:27.303 | INFO     | common.request_v1:wrapper:18 - --------请求
-------------
2024-01-06 23:45:27.303 | INFO     | common.request_v1:wrapper:24 - [method]: PUT [URL]: http://http***.org/put
2024-01-06 23:45:27.303 | DEBUG    | common.request_v1:wrapper:40 - [data]:
{
  "key": "value"
}
2024-01-06 23:45:27.978 | INFO     | common.request_v1:wrapper:48 - ---------响应
------------
2024-01-06 23:45:27.978 | INFO     | common.request_v1:wrapper:50 - successful with
```

```
status 200
2024-01-06 23:45:27.979 | DEBUG    | common.request_v1:wrapper:56 - [type]: json
[time]: 0.673543
2024-01-06 23:45:27.979 | DEBUG    | common.request_v1:wrapper:57 - [response]:
 {
  "args": {},
  "data": "",
  "files": {},
  "form": {
    "key": "value"
  },
  "headers": {
    "Accept": "*/*",
    "Accept-Encoding": "gzip, deflate",
    "Content-Length": "9",
    "Content-Type": "application/x-www-form-urlencoded",
    "Host": "httpbin.org",
    "User-Agent": "python-requests/2.31.0",
    "X-Amzn-Trace-Id": "Root=1-65997598-03ea159f2834214c79bfa00e"
  },
  "json": null,
  "origin": "203.176.241.236",
  "url": "http://http***.org/put"
}
2024-01-06 23:45:27.979 | INFO     | common.request_v1:wrapper:18 - ----------请求
-------------
2024-01-06 23:45:27.980 | INFO     | common.request_v1:wrapper:24 - [method]: DELETE
[URL]: http://http***.org/delete
2024-01-06 23:45:28.489 | INFO     | common.request_v1:wrapper:48 - ----------响应
-------------
2024-01-06 23:45:28.489 | INFO     | common.request_v1:wrapper:50 - successful with
status 200
2024-01-06 23:45:28.489 | DEBUG    | common.request_v1:wrapper:56 - [type]: json
[time]: 0.508469
2024-01-06 23:45:28.490 | DEBUG    | common.request_v1:wrapper:57 - [response]:
 {
  "args": {},
  "data": "",
  "files": {},
  "form": {},
  "headers": {
    "Accept": "*/*",
    "Accept-Encoding": "gzip, deflate",
    "Content-Length": "0",
    "Host": "httpbin.org",
    "User-Agent": "python-requests/2.31.0",
    "X-Amzn-Trace-Id": "Root=1-65997599-19ca90387e7ce9ab6656dc46"
  },
  "json": null,
  "origin": "203.176.241.236",
  "url": "http://http***.org/delete"
}
```

通过运行日志可以清晰地看到HTTP接口自动化测试的详细信息。

10.3 HTTP 接口测试断言设计

对于 HTTP 接口来说，大部分响应数据都为 JSON 格式。本节我们针对 JSON 格式的数据设计断言方法。

10.3.1 断言基础代码

一个比较常见的 HTTP 接口返回数据如下。

```
{
  "code": 100200,
  "message": "success",
  "result": {
    "user_list": [
      {
        "id": 1,
        "name": "tom",
        "hobby": [
          "basketball",
          "swimming"
        ]
      },
      {
        "id": 2,
        "name": "jack",
        "hobby": [
          "skiing",
          "reading"
        ]
      }
    ]
  }
}
```

在测试框架中编写测试断言时，需将 JSON 格式的字符串转换为字典（dict）格式，以便针对其中的每个字段进行数据提取及相应的断言操作，示例如下。

```
# 接口返回数据
resp = { ... }

# unittest 断言
self.assertEqual(resp["result"]["user_list"][0]["id"], 1)
self.assertEqual(resp["result"]["user_list"][0]["name"], "tom")
self.assertEqual(resp["result"]["user_list"][0]["hobby"][0], "basketball")
self.assertEqual(resp["result"]["user_list"][1]["name"], "jack")
self.assertEqual(resp["result"]["user_list"][1]["hobby"], ["skiing", "reading"])
```

基于 unittest 的断言方法来断言 JSON 格式的数据，与强类型语言相比，Python 在这方面做得比较简单，但是，对于复杂的 JSON 数据，显然还不够简捷。因此我们可以封装一些断言方法，以进一步简化对 JSON 格式的数据的断言。

📂 目录结构

```
10_chapter/
├──common
│   ├──__init__.py
│   ├──config.py
│   ├──case.py
│   └──request_v2.py
└──test_request_v2.py
```

config.py 文件的内容如下。

```python
class ResponseResult:
    """配置：响应结果"""
    status_code = 200
    response = None
```

📖 代码说明

定义 ResponseResult 类，status_code 变量用于全局存储 HTTP 状态码，response 变量用于全局存储 HTTP 的响应结果。

request_v2.py 文件的内容是在 10.2 节 request_v1.py 文件内容的基础上修改得来的。

```python
import json
import requests
from loguru import logger
from .config import ResponseResult

...

def request(func):
    """请求装饰器"""

    def wrapper(*args, **kwargs):
        func_name = func.__name__
        logger.info('-------------- 请求 -----------------')
        try:
            url = list(args)[1]
        except IndexError:
            url = kwargs.get("url", "")

        logger.info(f"[method]: {func_name.upper()}      [URL]: {url} ")
        auth = kwargs.get("auth", "")
        headers = kwargs.get("headers", "")
        cookies = kwargs.get("cookies", "")
        params = kwargs.get("params", "")
```

```python
            data = kwargs.get("data", "")
            json_ = kwargs.get("json", "")
            if auth != "":
                logger.debug(f"[auth]:\n {auth}")
            if headers != "":
                logger.debug(f"[headers]:\n {formatting(headers)}")
            if cookies != "":
                logger.debug(f"[cookies]:\n {formatting(cookies)}")
            if params != "":
                logger.debug(f"[params]:\n {formatting(params)}")
            if data != "":
                logger.debug(f"[data]:\n {formatting(data)}")
            if json_ != "":
                logger.debug(f"[json]:\n {formatting(json_)}")

            # running function
            r = func(*args, **kwargs)

            # 将HTTP的响应状态码赋值给ResponseResult.status_code变量
            ResponseResult.status_code = r.status_code

            logger.info("-------------- 响应 ----------------")
            if ResponseResult.status_code == 200 or ResponseResult.status_code == 304:
                logger.info(f"successful with status {ResponseResult.status_code}")
            else:
                logger.warning(f"unsuccessful with status {ResponseResult.status_code}")
            resp_time = r.elapsed.total_seconds()
            try:
                resp = r.json()
                logger.debug(f"[type]: json      [time]: {resp_time}")
                logger.debug(f"[response]:\n {formatting(resp)}")
                # 将HTTP的响应结果赋值给ResponseResult.response变量
                ResponseResult.response = resp
            except BaseException as msg:
                logger.debug("[warning]: failed to convert res to json, try to convert to text")
                logger.trace(f"[warning]: {msg}")
                r.encoding = 'utf-8'
                logger.debug(f"[type]: text      [time]: {resp_time}")
                logger.debug(f"[response]:\n {r.text}")
                # 将HTTP的响应结果赋值给ResponseResult.response变量
                ResponseResult.response = r.text

        return r

    return wrapper

...
```

📖 **代码说明**

与 request_v1.py 文件相比,request_v2.py 文件主要在 request 装饰器中增加了一些代码,将 HTTP 的响应状态码赋值给 config.py 文件的 ResponseResult.status_code 变量,将 HTTP 的响应结果赋值给 ResponseResult.response 变量。

case.py 文件的内容如下。

```python
import unittest
from .request_v2 import HttpRequest

# 定义 unittest 主方法
main = unittest.main


class TestCase(unittest.TestCase, HttpRequest):
    """
    定义 TestCase 类,使其同时继承 unittest.TestCase 类和 request-v2.py 文件的 HttpRequest 类
    """
    ...
```

📖 **代码说明**

创建 TestCase 类,使其同时继承 unittest.TestCase 类和 request_v2.py 文件的 HttpRequest 类。接下来,分别实现 assertPath()、assertJSON() 和 assertSchema() 断言方法。

10.3.2 assertPath()

JMESPath 是一种 JSON 查询语言,主要用于从 JSON 文档中提取和转换元素。

用 pip 命令安装 JMESPath。

```
> pip install jmespath
```

JMESPath 的基本用法如下。

```
>>> import jmespath
>>> data = {"error": {"code": 123, "msg": "error"}, "result": [{"id": 1, "name": "hello"}]}
>>> code = jmespath.search("error.code", data)
>>> code
123
>>> name = jmespath.search("result[0].name", data)
>>> name
'hello'
```

📂 **目录结构**

```
10_chapter/
```

```
├──common
│   ├──__init__.py
│   ├──config.py
│   ├──case.py
│   └──request_v2.py
└──test_request_v2.py
```

注意：本节的代码是在 10.3.1 节的项目代码基础上进行开发的。

修改 case.py 文件，增加 assertPath() 断言方法。

```python
import unittest
import jmespath
from loguru import logger
from.request_v2 import HttpRequest
from.config import ResponseResult

# 定义 unittest 主方法
main = unittest.main

class TestCase(unittest.TestCase, HttpRequest):
    """
    定义 TestCase 类，使其同时继承 unittest.TestCase 类和 request_v2.py 文件中的
HttpRequest 类
    """

    def assertPath(self, path: str, value: any) -> None:
        """
        断言 path 数据
        doc: https://jmesp***.org/
        :param path: JMESPath 的提取语法
        :param value: 断言值
        """
        logger.info(f"assertPath -> {path} >> {value}.")
        search_value = jmespath.search(path, ResponseResult.response)
        self.assertEqual(search_value, value)
```

📖 **代码说明**

首先，jmespath.search() 方法根据 path 语法提取 ResponseResult 类的 response 变量并获取响应数据。然后，调用 unittest 的 assertEqual() 断言方法判断是否等于预期值。

💡 **使用示例**

在 test_request_v2.py 文件中，编写测试用例并使用 assertPath() 断言方法。

```python
from common import case

class MyHttpTest(case.TestCase):
```

```python
    def test_assert_path(self):
        """
        测试 assertPath() 断言方法
        """
        params = {
            "user_list": [
                {
                    "id": 1,
                    "name": "tom",
                    "hobby": ["basketball", "swimming"]
                },
                {
                    "id": 2,
                    "name": "jack",
                    "hobby": ["skiing", "reading"]
                }
            ]
        }
        self.post("https://http***.org/post", json=params)
        self.assertPath("json.user_list[0].id", 1)
        self.assertPath("json.user_list[0].name", "tom")
        self.assertPath("json.user_list[0].hobby[0]", "basketball")
        self.assertPath("json.user_list[1].name", "jack")
        self.assertPath("json.user_list[1].hobby", ["skiing", "reading"])

if __name__ == '__main__':
    case.main()
```

当调用接口时，request 装饰器会将接口返回值保存到 ResponseResult 类的 response 变量中，assertPath()断言方法基于这个变量的值进行断言。因此，在使用 assertPath()断言方法时，只需要传入提取规则和预期的断言数据即可。

⏳ 运行结果

```
> python test_request_v2.py
2024-01-08 00:08:26.422 | INFO    | common.request_v2:wrapper:19 - --------- 请求 -------------
2024-01-08 00:08:26.422 | INFO    | common.request_v2:wrapper:25 - [method]: POST [URL]: https://http***.org/post
2024-01-08 00:08:26.423 | DEBUG   | common.request_v2:wrapper:43 - [json]:
 {
  "user_list": [
    {
      "id": 1,
      "name": "tom",
      "hobby": [
        "basketball",
        "swimming"
      ]
    },
```

```
      {
        "id": 2,
        "name": "jack",
        "hobby": [
          "skiing",
          "reading"
        ]
      }
    ]
  }
}
2024-01-08 00:08:27.614 | INFO    | common.request_v2:wrapper:51 - --------- 响应
------------
2024-01-08 00:08:27.614 | INFO    | common.request_v2:wrapper:53 - successful with status 200
2024-01-08 00:08:27.614 | DEBUG   | common.request_v2:wrapper:59 - [type]: json [time]: 1.188401
2024-01-08 00:08:27.615 | DEBUG   | common.request_v2:wrapper:60 - [response]:
 {
  "args": {},
  "data": "{\"user_list\": [{\"id\": 1, \"name\": \"tom\", \"hobby\": [\"basketball\", \"swimming\"]}, {\"id\": 2, \"name\": \"jack\", \"hobby\": [\"skiing\", \"reading\"]}]}",
  "files": {},
  "form": {},
  "headers": {
    "Accept": "*/*",
    "Accept-Encoding": "gzip, deflate",
    "Content-Length": "137",
    "Content-Type": "application/json",
    "Host": "httpbin.org",
    "User-Agent": "python-requests/2.31.0",
    "X-Amzn-Trace-Id": "Root=1-659acc7a-66b9f8bd2e48d5c427c097c8"
  },
  "json": {
    "user_list": [
      {
        "hobby": [
          "basketball",
          "swimming"
        ],
        "id": 1,
        "name": "tom"
      },
      {
        "hobby": [
          "skiing",
          "reading"
        ],
        "id": 2,
        "name": "jack"
      }
```

```
        ]
    },
    "origin": "203.176.241.236",
    "url": "https://http***.org/post"
}
2024-01-08 00:08:27.617 | INFO     | common.case:assertPath:26 - assertPath -> json.user_list[0].id >> 1.
2024-01-08 00:08:27.617 | INFO     | common.case:assertPath:26 - assertPath -> json.user_list[0].name >> tom.
2024-01-08 00:08:27.617 | INFO     | common.case:assertPath:26 - assertPath -> json.user_list[0].hobby[0] >> basketball.
2024-01-08 00:08:27.617 | INFO     | common.case:assertPath:26 - assertPath -> json.user_list[1].name >> jack.
2024-01-08 00:08:27.618 | INFO     | common.case:assertPath:26 - assertPath -> json.user_list[1].hobby >> ['skiing', 'reading']
```

使用JMESPath从JSON中提取元素显然要比使用字典（dict）更加简捷，下面是两种语法的对比。

```python
resp = self.post("https://http***.org/post", json=params)
data = resp.json()

self.assertEqual(data["json"]["user_list"][0]["id"], 1)
self.assertPath("json.user_list[0].id", 1)

self.assertEqual(data["json"]["user_list"][0]["name"], "tom")
self.assertPath("json.user_list[0].name", "tom")

self.assertEqual(data["json"]["user_list"][0]["hobby"][0], "basketball")
self.assertPath("json.user_list[0].hobby[0]", "basketball")

self.assertEqual(data["json"]["user_list"][1]["name"], "jack")
self.assertPath("json.user_list[1].name", "jack")

self.assertEqual(data["json"]["user_list"][1]["hobby"], ["skiing", "reading"])
self.assertPath("json.user_list[1].hobby", ["skiing", "reading"])
```

10.3.3　assertJSON()

如果要断言接口返回的整个或部分JSON数据，使用assertPath()断言方法将数据一个个提取出来断言就显得比较烦琐了，因此本节设计了assertJSON()断言方法，通过传入整个字典（dict）数据进行断言。

📂 目录结构

```
10_chapter/
├──common
│   ├──__init__.py
│   ├──utils.py
│   ├──config.py
```

```
    │   ├─case.py
    │   └─request_v2.py
    └─test_assert_v2.py
```

注意：本节的代码是在 10.3.1 节的项目代码基础上进行开发的，新增了 utils.py 文件。

首先，在 utils.py 文件中实现 diff_json()函数。

```python
class AssertInfo:
    """暂存断言信息"""
    warning = []
    error = []

def diff_json(response_data, assert_data):
    """
    递归：对比两个 JSON 格式的数据
    """

    if isinstance(response_data, dict) and isinstance(assert_data, dict):
        # 字典格式
        for key in assert_data:
            if key not in response_data:
                AssertInfo.error.append(f"错误: Response 没有 key: {key}")
        for key in response_data:
            if key in assert_data:
                # 递归
                diff_json(response_data[key], assert_data[key])
            else:
                AssertInfo.warning.append(f"警告: 断言数据没有 key: {key}")

    elif isinstance(response_data, list) and isinstance(assert_data, list):
        # 列表格式
        if len(response_data) == 0:
            AssertInfo.warning.append("警告: response 是[]")
        else:
            if isinstance(response_data[0], dict):
                try:
                    response_data = sorted(
                        response_data,
                        key=lambda x: x[list(response_data[0].keys())[0]])
                except TypeError:
                    response_data = response_data
            else:
                response_data = sorted(response_data)

        if len(response_data) != len(assert_data):
            AssertInfo.warning.append(
                f"警告: 列表长度不同: '{len(response_data)}' != '{len(assert_data)}'")

        if len(assert_data) > 0:
```

```
            if isinstance(assert_data[0], dict):
                try:
                    assert_data = sorted(
                        assert_data,
                        key=lambda x: x[list(assert_data[0].keys())[0]])
                except TypeError:
                    assert_data = assert_data
            else:
                assert_data = sorted(assert_data)

        for src_list, dst_list in zip(response_data, assert_data):
            # 递归
            diff_json(src_list, dst_list)
    else:
        if str(response_data) != str(assert_data):
            AssertInfo.error.append(f"错误：数据不相等：{response_data}")
```

📖 代码说明

创建 AssertInfo 类，warning 列表用于收集警告信息，error 列表用于收集错误信息。对二者的定义如下。

- warning：用于记录在返回数据（response_data）中有但在断言数据（assert_data）中没有定义的数据，并将其添加到警告列表中。因为在断言数据中没有定义，所以该数据"不重要"，可以不检查。

- error：用于记录在返回数据（response_data）中没有，但在断言数据（assert_data）中定义了的数据，并将其添加到错误列表中。因为在断言数据中有明确定义，所以说明该数据"非常重要"，必须要检查。

diff_json()是一个递归函数，用于接收返回数据和断言数据这两个参数。先对参数的类型进行判断，再根据参数类型进行对比。如果参数是 list 格式的，就通过 zip()方法将其打包为元组，之后循环调用 diff_json()函数；如果参数是 dict 格式的，就通过 sorted()方法进行排序，然后逐一进行对比。

在对比的过程中通过 warning 列表和 error 列表分别记录警告信息和错误信息。

接下来，在 case.py 文件中新增 assertJSON()断言方法。

```python
import unittest
from loguru import logger
from .request_v2 import HttpRequest
from .config import ResponseResult
from .utils import AssertInfo, diff_json

# 定义unittest主方法
main = unittest.main
```

```python
class TestCase(unittest.TestCase, HttpRequest):
    """
    定义 TestCase 类，继承 unittest.TestCase 和 HttpRequest
    """

    def assertJSON(self, assert_data, response=None) -> None:
        """
        断言 JSON 数据
        :param assert_data: JSON 数据
        :param response: 断言的 response，默认为 None
        """
        logger.info(f"assertJSON -> {assert_data}.")
        if response is None:
            response = ResponseResult.response

        AssertInfo.warning = []
        AssertInfo.error = []
        diff_json(response, assert_data)
        if len(AssertInfo.warning) != 0:
            logger.warning(AssertInfo.warning)
        if len(AssertInfo.error) != 0:
            self.assertEqual("Response data", "Assert data", msg=AssertInfo.error)
```

📖 代码说明

在 assertJSON() 断言方法中，assert_data 用于接收要断言的 dict 格式的数据，response 默认为 None，通过 ResponseResult.response 可获取接口响应数据。

首先，将 warning 列表和 error 列表的值置空，以免影响对后面数据的断言。

然后，调用 diff_json() 函数进行断言。如果 warning 列表长度不为 0，就通过 logger 打印警告信息；如果 error 列表长度不为 0，就说明测试用例失败，并打印错误信息。

🔧 使用示例

在 test_reques_v2.py 文件中编写测试用例，并使用 assertJSON() 断言方法。

```python
from common import case

class MyHttpTest(TestCase):

    def test_assert_json(self):
        """
        测试 assertJSON() 断言方法
        """
        payload = {"name": "tom", "hobby": ["basketball", "swim"]}
        resp = self.get("http://http***.org/get", params=payload)

        # 1.从整个 response 中断言
```

```python
        assert_data1 = {
            "args": {
                "hobby": ["swim", "basketball"],
                "name": "tom"
            }
        }
        self.assertJSON(assert_data1)

        # 2. 从部分 response 中断言
        assert_data2 = {
            "hobby": ["swim", "basketball"],
            "name": "tom"
        }
        self.assertJSON(assert_data2, resp.json()["args"])


if __name__ == '__main__':
    case.main()
```

📖 代码说明

assertJSON()断言方法的用法有两种。

- 用法 1：从整个 response 中断言，在 assert_data1 变量中定义断言的数据。对于不断言的字段可以不定义。
- 用法 2：从部分 response 中断言，在 assert_data2 变量中定义部分要断言的数据，需要填写第二个参数，指定 response 提取哪部分数据来断言。

⏳ 运行结果

```
> python test_reques_v2.py
2024-01-11 00:58:17.578 | INFO     | common.request_v2:wrapper:19 - ---------- 请求 ----------
2024-01-11 00:58:17.579 | INFO     | common.request_v2:wrapper:25 - [method]: GET       [URL]: http://http***.org/get
2024-01-11 00:58:17.579 | DEBUG    | common.request_v2:wrapper:39 - [params]:
{
  "name": "tom",
  "hobby": [
    "basketball",
    "swim"
  ]
}
successful with status 200
2024-01-11 00:58:18.225 | INFO     | common.request_v2:wrapper:51 - --------- 响应 -----------
2024-01-11 00:58:18.225 | INFO     | common.request_v2:wrapper:53 - 2024-01-11 00:58:18.225 | DEBUG    | common.request_v2:wrapper:59 - [type]: json       [time]: 0.63989
```

```
2024-01-11 00:58:18.225 | DEBUG   | common.request_v2:wrapper:60 - [response]:
{
  "args": {
    "hobby": [
      "basketball",
      "swim"
    ],
    "name": "tom"
  },
  "headers": {
    "Accept": "*/*",
    "Accept-Encoding": "gzip, deflate",
    "Host": "httpbin.org",
    "User-Agent": "python-requests/2.31.0",
    "X-Amzn-Trace-Id": "Root=1-659ecca8-02ebe87450c8659f0d54db25"
  },
  "origin": "203.176.241.236",
  "url": "http://http***.org/get?name=tom&hobby=basketball&hobby=swim"
}
2024-01-11 00:58:18.225 | INFO    | common.case:assertJSON:36 - assertJSON -> {'args':
{'hobby': ['swim', 'basketball'], 'name': 'tom'}}.
2024-01-11 00:58:18.226 | WARNING | common.case:assertJSON:44 - ['警告：断言数据没有
key: headers', '警告：断言数据没有 key: origin', '警告：断言数据没有 key: url']
2024-01-11 00:58:18.226 | INFO    | common.case:assertJSON:36 - assertJSON ->
{'hobby': ['swim', 'basketball'], 'name': 'tom'}.
.
----------------------------------------------------------------------
Ran 1 test in 0.565s

OK
```

运行日志可以更好地帮助我们理解两种用法的区别。用法 1 属于包含判断，会以警告的方式打印 response 中没有断言的数据。用法 2 属于相等判断，不会有告警，但如果两者不相等，则断言失败。

10.3.4　assertSchema()

在断言数据时，有时我们并不关心数据本身的值是什么，因为有些数据每次调用的值都不一样，比如，时间戳和 token。我们更关心数据的位置和数据的类型。比如，在下面的 JSON 格式的数据中，我们只关注 result 中是否有 key 为 id，且类型为 int 的数据即可。至于 id 的值是多少并不重要。以此类推，token 和 data_time 的值也只需类型是 string 即可。

```
{
  "result": {
    "id": 1,
    "token": "ahodfjasdfh1234h324kh2l3k",
    "data_time": "2022-11-12 12:00:00"
  }
```

}

JSON Schema 是一种声明性语言，可用于注释和验证 JSON 文档。它刚好可以用来解决上述问题。

用 pip 命令安装 jsonschema 库。

```
> pip install jsonschema
```

jsonschema 库的基本用法如下。

```
>>> from jsonschema import validate

>>> # 一个示例模式, 就像从 json.load() 函数中得到的那样
>>> schema = {
...     "type" : "object",
...     "properties" : {
...         "price" : {"type" : "number"},
...         "name" : {"type" : "string"},
...     },
... }

>>> # 如果 validate() 函数没有抛出异常, 则示例有效
>>> validate(instance={"name" : "Eggs", "price" : 34.99}, schema=schema)
```

📂 **目录结构**

```
10_chapter/
├──common
│  ├──__init__.py
│  ├──utils.py
│  ├──config.py
│  ├──case.py
│  └──request_v2.py
└──test_request_v2.py
```

注意：本节代码是在 10.3.1 节项目代码的基础上进行开发的。

在 case.py 文件中实现 assertSchema() 断言方法。

```python
import unittest
from loguru import logger
from jsonschema import validate
from jsonschema.exceptions import ValidationError
from .request_v2 import HttpRequest
from .config import ResponseResult


# 定义 unittest 主方法
main = unittest.main


class TestCase(unittest.TestCase, HttpRequest):
    """
```

定义 TestCase 类，使其同时继承 unittest.TestCase 类和 request-v2.Py 文件中的 HttpRequest 类
"""

```python
def assertSchema(self, schema, response=None) -> None:
    """
    Assert JSON Schema
    doc: https://json-sch***.org/
    :param schema: 断言的 schema 格式的数据
    :param response: 断言的 response，默认为 None
    """
    logger.info(f"assertSchema -> {schema}.")

    if response is None:
        response = ResponseResult.response

    try:
        validate(instance=response, schema=schema)
    except ValidationError as msg:
        self.assertEqual("Response data", "Schema data", msg)
```

📖 **代码说明**

在定义 assertSchema()断言方法时，第一个参数用于接收 schema 格式的数据；第二个参数为 response，默认为 None，即断言整个接口响应数据。

调用 jsonschema 提供的 validate()方法进行数据类型的判断。ValidationError 用于捕捉异常，如果出现异常，则以断言失败的方式抛出 msg 错误。

🔖 **使用示例**

在 test_request_v2.py 文件中创建测试用例，并使用 assertSchema()断言方法。

```python
from common import case


class MyHttpTest(case.TestCase):

    def test_assert_schema(self):
        """
        测试 assertSchema()断言方法
        """
        payload = {"hobby": ["basketball", "swim"], "name": "tom"}
        resp = self.get("http://http***.org/get", params=payload)

        # 1.从整个 response 中断言
        assert_data1 = {
            "type": "object",
            "properties": {
                "args": {
```

```python
                    "type": "object",
                    "properties": {
                        "hobby": {
                            "type": "array", "items": {"type": "string"}
                        },
                        "name": {
                            "type": "string"
                        }
                    }
                }
            }
        }
        self.assertSchema(assert_data1)

        # 2.从部分 response 中断言
        assert_data2 = {
            "type": "object",
            "properties": {
                "hobby": {
                    "type": "array", "items": {"type": "string"}
                },
                "name": {
                    "type": "string"
                }
            }
        }
        self.assertSchema(assert_data2, resp.json()["args"])


if __name__ == '__main__':
    case.main()
```

📖 代码说明

assert_data1 和 assert_data2 定义了要断言的数据结构和数据类型，type 用于指定类型，dict 格式的数据结构为 object，list 格式的数据结构为 array。数据的类型有 string、integer、boolean 等。这里的数据类型定义与 Python 中的有所不同。如果数据是 object 对象，则需要使用 properties 定义下一级数据。

assertSchema() 断言方法的用法分为两种。

- 用法 1：从整个 response 中断言，在 assert_data1 变量中定义完整的数据结构。
- 用法 2：从部分 response 中断言，在 assert_data2 变量中定义部分结构，并且需要填写第二个参数，指定 response 要断言的是哪部分数据。

⏳ 运行结果

```
> python test_request_v2.py
```

```
2024-01-15 23:33:51.710 | INFO    | common.request_v2:wrapper:19 - -------- 请求
--------------
2024-01-15 23:33:51.711 | INFO    | common.request_v2:wrapper:25 - [method]: GET
[URL]: http://http***.org/get
2024-01-15 23:33:51.712 | DEBUG   | common.request_v2:wrapper:39 - [params]:
{
  "hobby": [
    "basketball",
    "swim"
  ],
  "name": "tom"
}
2024-01-15 23:33:52.341 | INFO    | common.request_v2:wrapper:51 - ---------- 响应
------------
2024-01-15 23:33:52.343 | INFO    | common.request_v2:wrapper:53 - successful with
status 200
2024-01-15 23:33:52.344 | DEBUG   | common.request_v2:wrapper:59 - [type]: json
[time]: 0.625483
2024-01-15 23:33:52.345 | DEBUG   | common.request_v2:wrapper:60 - [response]:
{
  "args": {
    "hobby": [
      "basketball",
      "swim"
    ],
    "name": "tom"
  },
  "headers": {
    "Accept": "*/*",
    "Accept-Encoding": "gzip, deflate, br",
    "Host": "httpbin.org",
    "User-Agent": "python-requests/2.31.0",
    "X-Amzn-Trace-Id": "Root=1-65a55060-0be1613d04897ff3293d2f3c"
  },
  "origin": "223.197.61.61",
  "url": "http://http***.org/get?hobby=basketball&hobby=swim&name=tom"
}
2024-01-15 23:33:52.346 | INFO    | common.case:assertSchema:55 - assertSchema ->
{'type': 'object', 'properties': {'args': {'type': 'object', 'properties': {'hobby':
{'type': 'array', 'items': {'type': 'string'}}, 'name': {'type': 'string'}}}}}.
2024-01-15 23:33:52.350 | INFO    | common.case:assertSchema:55 - assertSchema ->
{'type': 'object', 'properties': {'hobby': {'type': 'array', 'items': {'type':
'string'}}, 'name': {'type': 'string'}}}.
.
----------------------------------------------------------------------
Ran 1 test in 0.644s

OK
```

10.4 实用功能封装

本节介绍接口测试中的实用功能封装。

10.4.1 HTTP 接口检查装饰器

在编写接口自动化测试过程中，除了接口测试，场景测试也非常重要。比如，电商网站的支付业务过程：

用户登录→浏览商品→加入购物车→下单→支付。

也就是说，如果想测试支付业务，就需要调用前面几个接口，因此我们需要对前面几个接口分别进行封装，以用户登录为例。

```python
import json
import requests

def get_token(username, password):
    """
    获取用户登录token
    :param username:
    :param password:
    """

    url = "http://http***.org/post"

    data = {
        "username": username,
        "password": password,
        "token": "token123"  # 假设这是接口返回的token
    }
    r = requests.post(url, data=data)

    if r.status_code != 200:
        raise ValueError("返回值请求失败")

    try:
        r.json()
    except json.decoder.JSONDecodeError:
        raise ValueError("返回值不是JSON格式")

    if r.json()["headers"]["Host"] != "httpbin.org":
        raise ValueError("接口返回的必要参数出现错误")

    try:
        user_token = r.json()["form"]["token"]
    except KeyError:
```

```
        user_token = ""

    return user_token

if __name__ == '__main__':
    token = get_token("admin", "abc123")
    print(f"get token: {token}")
```

📖 **代码说明**

get_token()方法用于获取用户登录 token，处理过程分为以下几个步骤。

步骤 1：判断状态码是否为 200。如果不是 200，则说明接口本身不通，抛出异常。

步骤 2：判断返回值格式是否为 JSON。如果不是，则可能接口返回数据格式错误，无法提取数据，抛出异常。

步骤 3：检查接口返回的必要参数，如 r.json()["headers"]["Host"]，如果不等于预期结果，就抛出异常。

步骤 4：提取接口返回的数据，如 r.json()["form"]["token"]，如果提取错误，就抛出异常或返回默认值。

我们在封装接口时，通常需要执行以上几个步骤。由于大部分接口都需要经过这些处理，我们完全可以通过实现一个装饰器来简化这一过程。

📁 **目录结构**

```
10_chapter/
├──tools
│   ├──__init__.py
│   └──response.py
└──test_response.py
```

创建 extend.py 文件，实现 check_response 装饰器。

```python
import json
from jmespath import search
from loguru import logger

def check_response(
        describe: str = "",
        status_code: int = 200,
        ret: str = None,
        check: dict = None,
        debug: bool = False):
    """
    检查返回的 response 数据
    :param describe: 接口描述
```

```python
        :param status_code: 断言状态码
        :param ret: 提取返回的数据
        :param check: 检查接口返回的数据
        :param debug: 日志开关 True/False
        :return:
        """
    def decorator(func):
        def wrapper(*args, **kwargs):
            func_name = func.__name__
            if debug is True:
                logger.info(f"Execute {func_name} - args: {args}")
                logger.info(f"Execute {func_name} - kwargs: {kwargs}")

            r = func(*args, **kwargs)
            flat = True
            if r.status_code != status_code:
                logger.info(f"Execute {func_name} - {describe} failed: {r.status_code}")
                flat = False

            try:
                r.json()
            except json.decoder.JSONDecodeError:
                logger.info(f"Execute {func_name} - {describe} failed: Not in JSON format")
                flat = False

            if debug is True:
                logger.info(f"Execute {func_name} - response:\n {r.json()}")

            if flat is True:
                logger.info(f"Execute {func_name} - {describe} success!")

            if check is not None:
                for expr, value in check.items():
                    data = search(expr, r.json())
                    if data != value:
                        logger.error(f"Execute {func_name} - check data failed: {value}")
                        raise ValueError(f"{data} != {value}")

            if ret is not None:
                data = search(ret, r.json())
                if data is None:
                    logger.warning(f"Execute {func_name} - return {ret} is None")
                return data
            else:
                return r.json()

        return wrapper
```

```
    return decorator
```

📖 **代码说明**

check_response 装饰器针对被装饰的函数和方法进行处理，所做的事情与 get_token()方法相似，只不过通过装饰器实现之后，可以在任意封装的接口方法中使用它。

需要注意的是，提取返回值的数据和返回值数据检查使用的都是 JMESPath 语言，详见 10.3.2 节，这里不赘述。

🐭 **使用示例**

```python
import requests
from common.extend import check_response

@check_response(
    describe="获取用户登录token",
    status_code=200,
    ret="form.token",
    check={"headers.Host": "httpbin.org"},
    debug=True)
def get_token(username, password):
    """
    获取用户登录token
    :param username:
    :param password:
    """
    url = "http://http***.org/post"
    data = {
        "username": username,
        "password": password,
        "token": "token123"
    }
    r = requests.post(url, data=data)
    return r

if __name__ == '__main__':
    token = get_token("jack", "pawd123")
    print(f"get token: {token}")
```

📖 **代码说明**

通过 check_response 装饰器接口调用函数，可以极大地减少样例代码的编写。

参数说明。

- describe：接口描述。
- status_code：检查接口返回值状态码是否为 200。

- ret：使用 JMESPath 提取器提取接口返回值中的 token。
- check：检查接口返回值中包含的参数，相当于对接口数据进行断言。
- debug：开启 debug，打印详细信息，以便调试。

⏳ 运行结果

```
> python user_response.py
2024-01-23 00:14:56.316 | INFO     | common.extend:wrapper:62 - Execute get_token - args: ('jack', 'pawd123')
2024-01-23 00:14:56.317 | INFO     | common.extend:wrapper:63 - Execute get_token - kwargs: {}
2024-01-23 00:14:56.776 | INFO     | common.extend:wrapper:78 - Execute get_token - response:
{'args': {}, 'data': '', 'files': {}, 'form': {'password': 'pawd123', 'token': 'token123', 'username': 'jack'}, 'headers': {'Accept': '*/*', 'Accept-Encoding': 'gzip, deflate', 'Content-Length': '45', 'Content-Type': 'application/x-www-form-urlencoded', 'Host': 'httpbin.org', 'User-Agent': 'python-requests/2.31.0', 'X-Amzn-Trace-Id': 'Root=1-65ae947f-5e18154e74fdbf18709d86c7'}, 'json': None, 'origin': '14.155.34.221', 'url': 'http://http***.org/post'}
2024-01-23 00:14:56.776 | INFO     | common.extend:wrapper:81 - Execute get_token - 获取用户登录token success!
get token: token123
```

10.4.2 方法依赖装饰器

在场景测试中，接口的依赖是不可避免的，下面我们简化这种依赖。

在执行测试用例过程中，需要知道完整的调用链。比如，测试用例依赖接口 B，接口 B 依赖接口 A。因此在调用过程中，要先调用接口 A，再调用接口 B。但是当系统十分复杂时，是没有人知道所有接口之间的调用依赖的。此时我们需要改变设计思路，对于每个接口，都只关心它的上一层级所依赖的接口，这样在编写接口场景测试用例时就简单多了。接口依赖如图 10-1 所示。

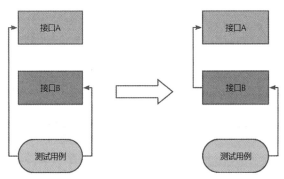

图 10-1 接口依赖

下面我们实现一个方法依赖装饰器来简化依赖调用。

📁 目录结构

```
10_chapter/
├──tools
│   ├──__init__.py
│   └──dependence.py
└──test_dependence.py
```

创建 dependence.py 文件，实现 dependent_func 装饰器。

```python
from typing import Callable, Text, Tuple
from functools import wraps
import redis
from loguru import logger

redis_client = redis.Redis(host='localhost', port=6379, db=0)

def dependent_func(func_obj: Callable, key_name: Text = None, *out_args,
**out_kwargs):
    """
    dependent_func装饰器
    :param func_obj: 方法对象
    :param key_name: 关键名
    :param out_args:
    :param out_kwargs:
    :return:
    """
    global redis_client
    def decorator(func):
        @wraps(func)
        def wrapper(*args, **kwargs):
            func_name = func.__name__
            depend_func_name = func_obj.__name__

            key = key_name
            if key_name is None:
                key = depend_func_name

            if not redis_client.get(key):
                dependence_res = _call_dependence(func_obj, func_name,*out_args,
**out_kwargs)
                redis_client.set(key, dependence_res)

            else:
                logger.info(f"<{depend_func_name}> 已被执行，通过缓存获取 `{key}`.")
            r = func(*args, **kwargs)
            return r

        return wrapper
```

```python
    return decorator

def _call_dependence(dependent_api: Callable or Text, func_name: Text, *args,
**kwargs):
    """
    执行依赖方法.
    :param dependent_api: 依赖方法
    :param func_name: 方法名
    :param args:
    :param kwargs:
    :return:
    """
    depend_func_name = dependent_api.__name__
    logger.info(f"<{func_name}> 依赖于 <{depend_func_name}>, 执行.")
    res = dependent_api(*args, **kwargs)
    return res
```

📖 代码说明

首先，dependent_func 装饰器依赖 Redis 缓存依赖方法执行的结果，所以需要安装 Redis 并启动 redis-server，同时需要安装 Python Redis 库，参考本书 7.3 节。

dependent_func 装饰器的参数：func_obj 参数用于接收被依赖的方法；key_name 用于指定被依赖的方法返回的数据在 Redis 中保存的 key。如果不指定，则默认使用被依赖的方法名作为 key；*out_args 和 **out_kwargs 用于指定被依赖的方法的参数。

dependent_func 装饰器的实现：通过 Redis 查询 key 是否存在，如果不存在，则调用 _call_dependence()方法获取被依赖的方法。

使用示例

下面基于前面的流程实现一个场景测试用例：

登录→获取项目 ID→基于项目 ID 创建项目模块。

```python
import hashlib
import unittest
from random import randint
import requests
from common.dependence import dependent_func, redis_client

class DependentTest(unittest.TestCase):

    @staticmethod
    def user_login(username, password):
        """
```

```python
        模拟用户登录，获取登录token
        """
        src = username + password
        m = hashlib.md5()
        m.update(src.encode())
        md5_str = m.hexdigest()
        print(f"生成token: {md5_str}")
        return md5_str

    @staticmethod
    @dependent_func(user_login, username="tom", password="t123")
    def get_project():
        """
        基于登录token，查询项目ID
        """
        token = redis_client.get("user_login")
        print(f"获取token: {token}，调用查询项目ID接口")
        pid = randint(1, 100)
        print(f"生成项目ID: {pid}")
        return pid

    @dependent_func(get_project, key_name="pid")
    def test_case(self):
        """
        测试用例，基于项目ID创建模块
        """
        pid = redis_client.get("pid")
        payload = {"pid": pid, "module_name": "module name"}
        r = requests.post("https://http***.org/post", data=payload)
        print(r.text)


if __name__ == '__main__':
    unittest.main()
```

📖 **代码说明**

user_login()方法用于接收username和password参数，模拟用户登录并获取token。为了简化代码，可以直接使用username和password生成一个md5值作为登录token。

get_project()方法用于获取项目ID，需要登录token才能查询项目ID。使用dependent_func装饰器调用user_login()方法，输入登录的username和password。同样为了简化代码，这里通过randint()函数生成一个随机数作为项目ID。

test_case()是最终要编写的测试用例，用于创建项目模块。首先，获取项目ID。使用dependent_func装饰器调用get_project()方法，通过key_name指定key的名称为pid。其次，通过Redis的get()方法，指定pid来获取get_project()方法，查询项目ID，调用项目ID接口运行创建模块。当然，这里的项目 ID 接口是使用 httpbin.org 的 API 模拟的。

虽然整个过程较为复杂，但是每个"接口"都只关心它自己所依赖的方法，之后用 dependent_func 装饰器运行被依赖的方法，通过 Redis 获取被依赖的方法的返回值。

⧗ 运行结果

```
> python test_dependence.py
2024-01-28 01:12:33.040 | INFO     | common.dependence:_call_dependence:53 - <test_case> 依赖于 <get_project>, 运行.
2024-01-28 01:12:33.041 | INFO     | common.dependence:_call_dependence:53 - <get_project> 依赖于 <user_login>, 运行.
生成token: 35e0ff9c4cba89998dda8255d0eb5408
获取token: b'35e0ff9c4cba89998dda8255d0eb5408', 调用项目ID接口
生成项目ID: 73
{
  "args": {},
  "data": "",
  "files": {},
  "form": {
    "module_name": "module name",
    "pid": "73"
  },
  "headers": {
    "Accept": "*/*",
    "Accept-Encoding": "gzip, deflate",
    "Content-Length": "30",
    "Content-Type": "application/x-www-form-urlencoded",
    "Host": "httpbin.org",
    "User-Agent": "python-requests/2.31.0",
    "X-Amzn-Trace-Id": "Root=1-65b53981-794cc9dc71f1e70c0ee9120e"
  },
  "json": null,
  "origin": "203.176.241.236",
  "url": "https://http***.org/post"
}

.
----------------------------------------------------------------------
Ran 1 test in 3.331s

OK
```

通过运行结果，我们可以更好地理解依赖方法的调用过程。

dependent_func 装饰器可以同时依赖 A 和 B 两个方法，用法如下。

```
import unittest
from common.dependence import redis_client, dependent_func
```

```
class DependentTest(unittest.TestCase):

    def get_name(self):
        return "tom"

    def get_age(self):
        return 23

    @dependent_func(get_name, key_name="name")
    @dependent_func(get_age, key_name="age")
    def test_case(self):
        """
        sample test case
        """
        name = redis_client.get("name")
        age = redis_client.get("age")
        ...
```

除此之外，dependent_func 装饰器拥有更多的用法，支持更为复杂的场景，读者可自行探索。

10.4.3 生成 curl 命令

我们可以在任意终端执行 curl 命令，当然，前提是需要安装它。curl 命令的应用非常广泛，我们可以在浏览器的"开发者工具"中直接复制接口，生成 curl 命令，如图 10-2 所示。

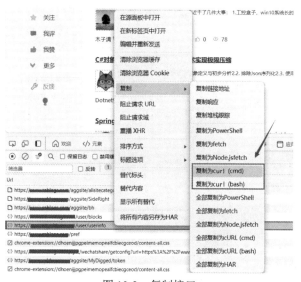

图 10-2 复制接口

我们可以将 requests 库的请求对象转为 curl 命令，以便在日常开发中将其转换为其他工具。比如，在终端执行 curl 命令，或者将 curl 命令导入 postman 等接口工具。

当我们使用 requests 库实现一个接口调用时，request 对象包含了请求的 method、headers、url 和 body 等数据。

```python
# requests_demo2.py
import requests

r = requests.get("https://http***.org/post", json={"key": "value"},
headers={"token": "abc123"})
print(r.request.method)
print(r.request.headers)
print(r.request.url)
print(r.request.body)
```

⏳ **运行结果**

```
> python requests_demo2.py
GET
{'User-Agent': 'python-requests/2.31.0', 'Accept-Encoding': 'gzip, deflate',
'Accept': '*/*', 'Connection': 'keep-alive', 'token': 'abc123', 'Content-Length':
'16', 'Content-Type': 'application/json'}
https://http***.org/post
b'{"key": "value"}'
```

接下来，将 requests 库的请求对象转换为 curl 命令进行封装。

📁 **目录结构**

```
10_chapter/
├──tools
│   ├──__init__.py
│   └──curl.py
└──test_curl.py
```

创建 curl.py 文件，实现 to_curl() 方法。

```python
from shlex import quote

def to_curl(request) -> str:
    """
    将 request 对象转换为 curl 命令
    :param request: request 对象
    """
    parts = [
        ('curl', None),
        ('-X', request.method),
    ]

    for key, value in sorted(request.headers.items()):
        parts += [('-H', f'{key}: {value}')]

    if request.body:
        body = request.body
```

```
        if isinstance(body, bytes):
            body = body.decode('utf-8')
        parts += [('-d', body)]

    parts += [(None, request.url)]

    flat_parts = []
    for key, value in parts:
        if key:
            flat_parts.append(quote(key))
        if value:
            flat_parts.append(quote(value))

    return ' '.join(flat_parts)
```

📖 代码说明

实现 to_curl() 方法，将传入的 request 对象转换为 curl 命令。

- -X 参数指定了请求方法。
- -H 参数指定了请求头。
- -d 参数指定了 body 参数。

🖰 使用示例

下面通过示例演示 to_curl() 方法的使用。

```
import requests
from common.curl import to_curl

def test_to_curl():
    """
    转换为 curl 命令
    """
    r = requests.get('http://http***.org/get', params={'key': 'value'})
    curl_command = to_curl(r.request)
    print(curl_command)

    r = requests.post('http://http***.org/post', data={'key': 'value'})
    curl_command = to_curl(r.request)
    print(curl_command)

    r = requests.delete('http://http***.org/delete', json={'key': 'value'})
    curl_command = to_curl(r.request)
    print(curl_command)

    r = requests.put('http://http***.org/put', json={'key': 'value'},
headers={"token": "123"})
    curl_command = to_curl(r.request)
    print(curl_command)
```

```
if __name__ == '__main__':
    test_to_curl()
```

将 GET、POST、DELETE 和 PUT 方法转换为 curl 命令。

⌛ 运行结果

> python test_to_curl.py

curl -X GET -H 'Accept: */*' -H 'Accept-Encoding: gzip, deflate' -H 'Connection: keep-alive' -H 'User-Agent: python-requests/2.31.0' 'http://http***.org/get?key=value'

curl -X POST -H 'Accept: */*' -H 'Accept-Encoding: gzip, deflate' -H 'Connection: keep-alive' -H 'Content-Length: 9' -H 'Content-Type: application/x-www-form-urlencoded' -H 'User-Agent: python-requests/2.31.0' -d key=value http://http***.org/post

curl -X DELETE -H 'Accept: */*' -H 'Accept-Encoding: gzip, deflate' -H 'Connection: keep-alive' -H 'Content-Length: 16' -H 'Content-Type: application/json' -H 'User-Agent: python-requests/2.31.0' -d '{"key": "value"}' http://http***.org/delete

curl -X PUT -H 'Accept: */*' -H 'Accept-Encoding: gzip, deflate' -H 'Connection: keep-alive' -H 'Content-Length: 16' -H 'Content-Type: application/json' -H 'User-Agent: python-requests/2.31.0' -H 'token: 123' -d '{"key": "value"}' http://http***.org/put

我们可以将上面生成的 curl 命令复制到终端并执行，如图 10-3 所示。

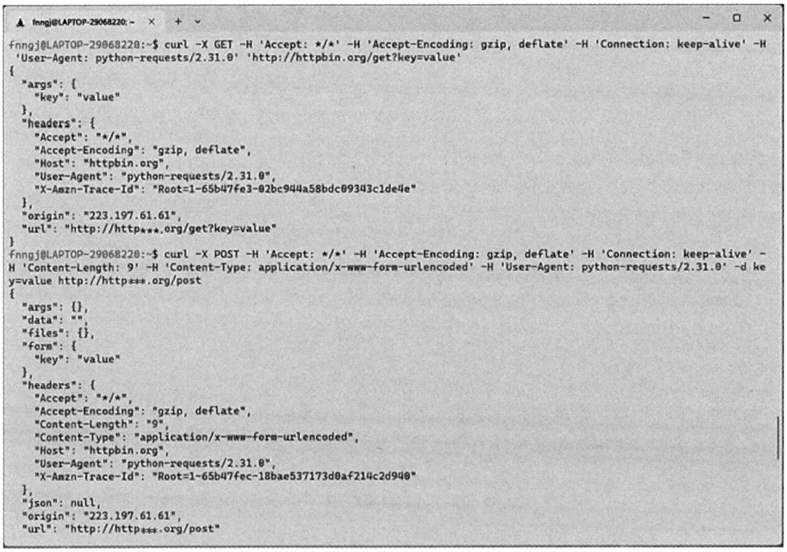

图 10-3　将生成的 curl 命令复制到终端并执行

10.5　WebSocket 封装与测试

WebSocket 是一种网络通信协议，它支持在单个长连接上进行全双工、双向交互。它使得数据既可以从客户端传送到服务端，也可以从服务端传送到客户端，并且这种通信是实时的，非常适合需要频繁和低延迟交互的应用场景。因此，WebSocket 被广泛应用于在线游戏、实时通信和实时数据监控等场景。

10.5.1　WebSocket 封装

websocket-client 是 Python 的一个 WebSocket 客户端库，它提供了对 WebSocket 低级 API 的访问功能。我们可以基于 websocket-client 库封装一个 WebSocket 调用客户端。

用 pip 命令安装 websocket-client 库。

```
> pip install websocket-client
```

📂 目录结构

```
10_chapter/
├──tools
│   ├──__init__.py
│   └──websocket_client.py
└──test_websocket.py
```

接下来，基于 websocket-client 库和多线程，封装一个 WebSocket 调用客户端。

```python
from threading import Thread
import websocket

class WebSocketClient(Thread):
    """
    WebSocket Client
    """

    def __init__(self, url):
        Thread.__init__(self)
        self.ws = websocket.create_connection(url)
        self.running = True
        self.received_messages = []

    def run(self):
        """
        实现运行方法
        :return:
        """
        while self.running:
            try:
                message = self.ws.recv()
```

```
                self.received_messages.append(message)
            except websocket.WebSocketConnectionClosedException:
                break
            except Exception as e:
                print(f"WebSocket error: {e}")
                break
        self.ws.close()

    def send_message(self, message):
        """
        发送消息
        :param message:
        :return:
        """
        self.ws.send(message)

    def stop(self):
        """
        停止运行
        :return:
        """
        self.running = False
```

📖 **代码说明**

创建 WebSocketClient 类，使其集成 Thread 线程类。在 __init__()初始化方法中，通过 websocket（即 websocket-client）提供的 create_connection()方法创建 WebSocket 连接，并赋值给 ws 变量；running 变量用于控制是否关闭连接；received_messages 列表用于记录发送的消息。

重写 Thread 类的 run()方法，用 while 循环判断 running 变量是否为 True, 持续监听 WebSocket 消息，并记录到 received_messages 列表。在跳出循环后，通过 ws.close()函数关闭 WebSocket 连接。

实现 send_message()方法，通过调用 ws.send()函数发送 WebSocket 消息。

实现 stop()方法，该方法将 running 变量设置为 False，即跳出 while 循环，停止执行。

10.5.2　WebSocket 测试

参考 10.1.3 节中的内容，通过 aiohttp 库实现 WebSocket 接口，重新启动服务。

```
> python aiohttp_server_demo.py
======== Running on http://127.0.0.1:8080 ========
(Press CTRL+C to quit)
```

下面基于 unittest 编写 WebSocket 测试用例。

```
import time
import unittest
```

```python
from tools.webcocket_client import WebSocketClient

class WebSocketTest(unittest.TestCase):
    def setUp(self):
        """
        创建 WebSocket 客户端线程
        """
        self.client = WebSocketClient("ws://127.0.0.1:8080/echo")
        self.client.start()
        # 等待客户端连接建立
        time.sleep(1)

    def tearDown(self):
        """
        发送关闭消息
        """
        self.client.send_message("close")
        # 停止 WebSocket 客户端线程
        self.client.stop()
        self.client.join()

    def test_send_and_receive_message(self):
        """
        测试发送消息
        """
        self.client.send_message("WebSocket!")
        time.sleep(1)
        print(self.client.received_messages)
        self.assertEqual(len(self.client.received_messages), 1)
        self.assertIn("Hello, WebSocket!", self.client.received_messages[0])

if __name__ == '__main__':
    unittest.main()
```

📖 **代码说明**

setUp()方法用于测试用例开始之前的准备工作。在 setUp()方法中调用 WebSocketClient()函数创建 WebSocket 连接，并将连接对象赋值给 self.client 变量，通过 start()方法激活客户端线程。

tearDown()方法用于测试用例运行完毕之后的清理工作。在 tearDown()方法中，通过调用 client.send_message()方法发送"close"命令，主动请求关闭 WebSocket 连接。这取决于 WebSocket 服务端是如何设计的，我们在 WebSocket 服务端实现了下面的逻辑。

```
if msg.data == 'close':
    await ws.close()
```

如果接收的消息为 close，则关闭 WebSocket 连接。client.stop()方法用于将 running 变量设置为 False，client.join()方法用于等待线程结束。

test_send_and_receive_message()方法用于编写测试用例。client.send_message()方法用于向服务端发起请求。client.received_message 参数用于记录发送了哪些请求，我们可以基于其中的数据进行断言。

运行结果

```
> python test_websocket.py
['Hello, WebSocket!']
.
----------------------------------------------------------------------
Ran 1 test in 2.009s

OK
```

第11章 自动化测试设计模式

在软件开发过程中,我们会用到许多设计模式,比如,单例模式、工厂模式和观察者模式。在软件自动化测试过程中,同样有一些常见的设计模式,比如,Page Object 模式、Bot 模式和 API Object 模式,本章将深入讲解这些模式的应用。

11.1 设计模式与开发策略

本节我们主要讲解在 UI 自动化测试中常用的设计模式和开发策略。

📂 目录结构

```
11_chapter/
├──design_patterns
│   ├──__init__.py
│   ├──bot_style.py
│   └──page_object.py
├──test_bot_style.py
└──test_page_object.py
```

11.1.1 Page Object 模式

Page Object 模式在 Web UI 和 App UI 自动化测试领域极为流行,其核心宗旨是增强测试代码的可维护性,并有效减少代码重复。

下面简单封装 Bing 搜索中的相关元素定位,创建一个名为 page_object.py 的文件。

```
# page_object.py
from selenium.webdriver.remote.webdriver import WebDriver

class BingPage:
```

```python
    def __init__(self, driver: WebDriver):
        self.driver = driver

    @property
    def search_input(self):
        return self.driver.find_element("id", "sb_form_q")

    @property
    def search_icon(self):
        return self.driver.find_element("id", "search_icon")
```

📖 **代码说明**

创建 BingPage 类，在 __init__()初始化方法中接收驱动对象；之后，将 Bing 首页的搜索输入框和 icon 的元素定位封装成方法。因为元素本身不需要传参，所以可以使用 property 装饰器将方法转变为只读属性。

💡 **使用示例**

调用 BingPage 类编写自动化脚本。

```python
from time import sleep
from selenium.webdriver import Chrome
from design_patterns.page_object import BingPage

driver = Chrome()
driver.get("https://cn.b***.com")

# 调用 BingPage 类
page = BingPage(driver)
page.search_input.send_keys("page object tests")
page.search_icon.click()

driver.close()
```

首先，调用 BingPage 类，传入浏览器驱动 driver。然后针对 BingPage 类中定义的元素对象进行 send_keys 和 click 操作。

11.1.2 Bot 模式

尽管 Page Object 模式可以有效减少代码重复，但它并不总是团队愿意遵循的模式。另一种方法是遵循 command-like 风格的测试。

Bot 是基于原始 Selenium API 的面向操作的抽象。如果我们发现命令没有为应用做正确的事情，则可以很容易地修改它们。

封装元素的输入和点击操作，创建一个名为 bot_style.py 的文件。

```python
# bot_style.py
from selenium.webdriver.remote.webdriver import WebDriver

class ActionBot:

    def __init__(self, driver: WebDriver):
        self.driver = driver

    def click(self, *locator, times=1):
        """
        点击
        :param locator: 元素定位
        :param times: 点击次数
        :return:
        """
        elem = self.driver.find_element(*locator)
        for _ in range(times):
            elem.click()

    def type(self, *locator, text):
        """
        输入
        :param locator: 元素定位
        :param text: 输入文本
        :return:
        """
        elem = self.driver.find_element(*locator)
        elem.clear()
        elem.send_keys(text)
```

📖 代码说明

这种设计更倾向于封装 Selenium API，即它不关心元素定位的细节，只聚焦于动作的封装。比如，在 click() 方法中，可以通过 times 参数来控制点击次数；在 type() 方法中，可以集成 clear() 方法，以便在输入内容前自动清空输入框。

🖐 使用示例

调用 ActionBot 类，编写自动化测试脚本。

```python
from selenium.webdriver import Chrome
from selenium.webdriver.common.by import By
from design_patterns.bot_style import ActionBot

driver = Chrome()
driver.get("https://cn.b***.com")

# 调用 ActionBot 类
```

```
action_bot = ActionBot(driver)
action_bot.type(By.ID, "sb_form_q", text="bot style tests")
action_bot.click(By.ID, "search_icon", times=1)

driver.quit()
```

首先,调用 ActionBot 类,传入浏览器驱动 driver。然后,调用 ActionBot 类下面的 type() 和 click() 方法进行输入和点击操作,当然,我们需要将元素定位作为参数传入。

Page Object 模式和 Bot 模式之间并无优劣之分,它们只是一种编码风格,并不会对代码的最终执行结果产生影响,我们在工作中可以结合项目选择合适的设计模式。

11.2 基于 Page Object 模式的相关库

在深入理解 Page Object 模式的设计原理之后,采用何种编码方式都是可行的。除了直接在页面对象层(Page Layer)编写元素定位代码,我们还可以选择使用一些现成的库来简化页面对象层的代码编写工作。这些库能够帮助我们以更高效、更可维护的方式实现页面元素的定位和操作。

> 目录结构

```
11_chapter/
├── page_object_lib
│   ├── __init__.py
│   ├── spf_page.py
│   └── poium_page.py
├── test_spf_demo.py
└── test_poium_demo.py
```

11.2.1 selenium-page-factory

selenium-page-factory 是一个 Python 库,它通过提供 PageFactory 类,使得在 Selenium 自动化测试中实现 Page Object 模式变得更加便捷。它的特点如下。

- 一次性初始化 Point 中声明的所有 WebElement。
- 所有的 WebElement 都经过重新定义,以引入额外的功能。比如,click()方法已经被扩展,加入了显式等待的机制,确保元素在被点击之前是可点击的。
- 支持 Selenium 4 ActionChains()方法。
- 支持 Appium 移动测试。
- 自定义 Page Factory 异常。

用 pip 命令安装 selenium-page-factory。

```
> pip install selenium-page-factory
```

使用示例

基于 selenium-page-factory 实现对 Page 层的封装。

```python
# spf_page.py
from seleniumpagefactory.Pagefactory import PageFactory

class BingPage(PageFactory):

    def __init__(self, driver):
        self.driver = driver
        self.timeout = 15
        self.highlight = True

    # 使用 PageFactory 类定义定位器字典
    locators = {
        "searchInput": ('ID', 'sb_form_q'),
        "searchIcon": ('ID', 'search_icon'),
    }

    def search(self, text):
        """
        Bing search
        """
        self.searchInput.set_text(text)
        self.searchIcon.click_button()
```

代码说明

创建 BingPage 类，使其继承 PageFactory 类，重写父类的 __init__()初始化方法，定义 driver、timeout（超时时间），以及 highlight（操作元素高亮）。

locators 用于定义页面中元素的定位。

定义 search()方法，用于封装业务动作。定义元素的文本输入 set_text()方法和点击按钮 click_button()方法。set_text()方法和 click_button()方法都是 PageFactory 类提供的扩展方法。

使用示例

调用 BingPage 层的封装，编写自动化测试脚本。

```python
from selenium import webdriver
from page_object_lib.spf_page import BingPage

driver = webdriver.Chrome()
driver.get("https://cn.b***.com")

bp = BingPage(driver)
bp.search("selenium-page-factory")
```

```
driver.quit()
```

首先，调用 BingPage 类传参浏览器驱动。然后，调用 search() 方法传参搜索的关键字。

在使用 selenium-page-factory 的过程中发现以下两个问题。

- 没有提供对一个元素进行返回与操作的功能。比如，只能返回元素列表，或者定位方法只能定位到一组元素，无法进一步通过索引指定是第几个元素。
- 对 Page 层的定义过重，既包含元素定位，又包含业务操作逻辑。在测试脚本层面，只需调用 search() 方法即可，这是 selenium-page-factory 推荐的使用方式。虽然这样的设计简化了自动化脚本层面的操作，但是也牺牲了一定的灵活性。

11.2.2　poium 的基本使用

poium 是一个基于 Page Object 模式的测试库，它同时支持 Appium、Selenium、Playwright、UiAutomator2 和 facebook-wda 等 Web 和 App UI 测试库，其主要特点如下。

- 以极简方式定义 Page 层元素。
- 支持主流的 Web 和 App UI 测试库。
- 对原生 API 无损。

用 pip 命令安装 poium。

```
> pip install poium
```

🔰 使用示例

基于 poium 实现对 Page 层的封装。

```python
# poium_page.py
from poium import Page, Element, Elements

class BingPage(Page):
    """Bing 搜索页面"""
    search_input = Element("#sb_form_q", describe="Bing 搜索输入框")
    search_icon = Element("#search_icon", describe="Bing 搜索输入框")
    search_result = Elements("//h2/a", describe="搜索结果", timeout=3)

class CalculatorPage(Page):
    """手机默认自带计算器"""
    number_1 = Element(id_="com.huawei.calculator:id/digit_1")
    number_2 = Element(id_="com.huawei.calculator:id/digit_2")
    add = Element(id_="com.huawei.calculator:id/op_add")
    eq = Element(id_="com.huawei.calculator:id/eq")
```

📖 代码说明

创建一个名为 BingPage 的类，该类继承自 Page 类。在这个类中，我们定义了 Element 类和 Elements 类，分别用于定位页面上的单个元素和一组元素。describe 参数用于定义元素的描述信息，而 timeout 参数则用于设置元素定位的超时时间，默认为 5s。

Element 类和 Elements 类的定位方法与 Selenium 和 Appium 的基本保持一致。

🔧 使用示例

对于 Web 自动化测试，我们通过调用 BingPage 层的封装方法来实现对 Bing 搜索引擎的操作，实现 Bing 搜索，并打印搜索结果列表。

```
from selenium import webdriver
from page_object_lib.poium_page import BingPage

driver = webdriver.Chrome()
driver.get("https://cn.b***.com")

bp = BingPage(driver)
bp.search_input.send_keys("poium")
bp.search_icon.click()
bp.sleep(2)

result = bp.search_result
for r in result:
    print("搜索结果：", r.text)
```

对于 App 自动化测试，我们通过调用 CalculatorPage 层的封装方法来实现加法操作。

```
from appium import webdriver
from appium.options.android import UiAutomator2Options
from page_object_lib.poium_page import CalculatorPage

capabilities = {
    "automationName": "UiAutomator2",
    "platformName": "Android",
    'appPackage': 'com.huawei.calculator',
    'appActivity': '.Calculator',
}
options = UiAutomator2Options().load_capabilities(capabilities)
driver = webdriver.Remote('http://127.0.0.1:4723', options=options)

page = CalculatorPage(driver)
page.number_1.click()
page.add.click()
page.number_2.click()
page.eq.click()
```

⌛ 运行结果

```
> python test_poium_web.py
2024-02-13 21:40:38 logging.py | INFO | 🔍 Find element: id=sb_form_q. Bing 搜索输入框
2024-02-13 21:40:39 logging.py | INFO | ☑ send_keys('poium').
2024-02-13 21:40:39 logging.py | INFO | 🔍 Find element: id=search_icon. Bing 搜索输入框
2024-02-13 21:40:41 logging.py | INFO | ☑ click().
2024-02-13 21:40:43 logging.py | INFO | ✨ Find 11 elements through: xpath=//h2/a. 搜索结果
搜索结果: GitHub - SeldomQA/poium: Page Objects design pattern …
搜索结果: 【unittest 单元测试框架】（10）poium 测试库 - hello_般 - 博客园
搜索结果: poium · PyPI
搜索结果: poium - 必应词典
搜索结果: pytest 自动化测试中的 poium 测试库_python poium-CSDN 博客
搜索结果: poium: Selenium/appium-based Page Objects test library
搜索结果: poium · TesterHome
搜索结果: poium/README.md at master · SeldomQA/poium · GitHub
搜索结果: Poium 测试库在 pytest 自动化测试中的应用-百度开发者中心
搜索结果: poium 1.2.0 on PyPI - Libraries.io

> python test_poium_app.py
2024-02-13 21:43:45 logging.py | INFO | 🔍 Find element: id=com.huawei.calculator:id/digit_1.
2024-02-13 21:43:45 logging.py | INFO | ☑ click().
2024-02-13 21:43:45 logging.py | INFO | 🔍 Find element: id=com.huawei.calculator:id/op_add.
2024-02-13 21:43:46 logging.py | INFO | ☑ click().
2024-02-13 21:43:46 logging.py | INFO | 🔍 Find element: id=com.huawei.calculator:id/digit_2.
2024-02-13 21:43:46 logging.py | INFO | ☑ click().
2024-02-13 21:43:47 logging.py | INFO | 🔍 Find element: id=com.huawei.calculator:id/eq.
2024-02-13 21:43:47 logging.py | INFO | ☑ click().
```

运行结果涵盖了 Web 和 App 两个自动化测试脚本。

在 Web 自动化测试中，poium 支持元素高亮显示功能，使我们更容易观察页面当前操作的元素。此外，它提供了日志记录功能，让我们能够追踪整个测试执行过程。

poium 被认为对原生 API 无损，是因为它避免了"过度封装"，仅仅负责简化元素的定义，不干预元素的具体动作或动作组合。在编写自动化测试脚本时，我们可以根据需要有选择地使用 poium，无论是全面采用还是仅用于部分元素的定义，其应用方式都极为灵活。

11.2.3　poium 的设计原理

poium 因其简单易用的特性而广受欢迎。它的设计原理十分简洁，主要利用了 Python 语言的语法糖特性。理论上，poium 能够与任何 UI 测试库兼容，提供广泛的支持。

本节我们实现 poium 的最小模型。

📂 **目录结构**

```
11_chapter/
├──poium_core
│   ├── __init__.py
│   └──page.py
└──test_poium_core.py
```

下面实现 poium 的核心代码。

```python
# poium_core/page.py

class Page:
    """
    定义 Page 类
    """
    def __init__(self, driver):
        self.driver = driver


class Element:
    """
    定义 Element 类
    """
    def __init__(self, *locator):
        if len(locator) < 1:
            raise ValueError("Please specify only one locator")
        self.locator = locator

    def __get__(self, instance, owner):
        if instance is None:
            return None
        self.driver = instance.driver
        return self

    def input(self, text: str):
        """
        input 输入
        :param text: 输入文本
        """
        self.driver.find_element(*self.locator).send_keys(text)

    def click(self):
        """
```

```
click 点击
"""
self.driver.find_element(*self.locator).click()
```

📖 **代码说明**

首先，创建 Page 类，在 __init__()初始化方法中通过 driver 参数接收浏览器驱动对象，从而初始化页面对象。

其次，创建 Element 类，在 __init__()初始化方法中接收一个包含元素定位信息的*locator 参数，用于确定页面上特定元素的位置。

__get__()方法是 Python 的一个内置方法。当一个类定义了这个方法时，它可以控制属性的访问方式。如果通过一个实例访问这个属性，那么 instance 参数将指向该实例。如果通过类本身访问这个属性，那么 instance 参数将为 None。owner 参数始终指向 Owner 类。

在 Element 类中，通过 instance.driver 变量获取的驱动对象实际上是从 Page 类的 self.driver 变量继承而来的，这与 Page 类和 Element 类的使用方式密切相关。

input()方法和 click()方法在获取 self.driver 变量之后，对元素进行定位和操作。

扩展知识：

在 Python 中，一个类只要实现了__get__()方法、__set__()方法或__delete__()方法中的任意一个，就可以称它为描述器（descriptor）。如果类只定义了__get__()方法，则称它为非资料描述器（non-data descriptor）。如果类定义了__set__()或__delete__()方法中的任意一个，或者两者都定义了，则称它为资料描述器（data descriptor）。

🔖 **使用示例**

下面通过示例演示 Page 类和 Element 类的使用。

```python
from selenium.webdriver import Chrome
from selenium.webdriver.common.by import By
from poium_core.page import Page, Element

class BingPage(Page):
    """Bing 页面元素"""
    search_input = Element(By.ID, "sb_form_q")
    search_icon = Element(By.ID, "search_icon")


if __name__ == '__main__':
    driver = Chrome()
    driver.get("https://cn.b***.com")
    bp = BingPage(driver)
```

```
bp.search_input.input("poium")
bp.search_icon.click()
```

创建一个名为 BingPage 的类，该类继承自 Page 类。这意味着在 BingPage 类的 __init__() 方法中，我们可以指定浏览器驱动，从而初始化页面对象。Element 类作为 BingPage 类的一个属性，允许我们通过实例（instance 参数）访问 BingPage 类的 self.driver 变量。

这种设计十分巧妙，一旦理解了这个概念，poium 的使用就变得不再神秘。此外，这个知识点的应用非常灵活，我们可以将其应用到其他场景，实现举一反三的效果。

11.3 API Object 模式

API Object 模式简称为 AOM。它专注于提升 API 测试和集成的直观性和灵活性。通过将 API 封装成易于操作的对象，AOM 能够有效地简化用户在使用 API 时面临的复杂性，如请求处理、响应解析、异常管理和身份验证等。这种封装方式使用户能够更加专注于业务逻辑，而不必深陷 API 的具体实现细节。

11.3.1 AOM 的设计原理

1. AOM 的基本用法

我们可以将业务高度关联的一组 API 封装为一个 API Object。比如，在一个购物网站的 API 测试中，可以设计一个名为 "OrderAPIObject" 的对象，用以抽象化订单处理流程中的复杂性。这个对象整合了添加商品至购物车、设置收件信息和提交订单等所需的 API 调用。测试脚本只需与 "OrderAPIObject" 对象进行交互即可。这样不仅简化了测试流程，也使得测试过程更加直观和易于管理。

```
class OrderAPIObject:

    def add_item_to_cart(self, item_id):
        """
        发出 API 请求，添加商品至购物车
        :param item_id:
        :return:
        """
        ...

    def set_shipping_details(self, details):
        """
        通过 API 请求设置收件信息
        :param details:
        :return:
        """
```

```
    ...
    def place_order(self):
        """
        提交订单并接收确认
        :return:
        """
        ...
```

2. 创建 API 对象的技巧

为了构建一个功能强大的 API 对象，我们对"OrderAPIObject"对象进行了扩展。在下单流程中调用 prepare_order()方法，用以执行下单前所需的前置工作。此外，place_order()方法可以集成异常处理机制，以及对错误状态码的响应处理，从而确保测试过程保持弹性。

```
import requests

class OrderAPIObject:

    def __init__(self):
        # 调用prepare_order()方法
        self.prepare_order()

    def prepare_order(self):
        """
        准备下单前所需的前置工作
        :return:
        """
        ...

    def place_order(self) -> dict:
        """
        下单并处理错误响应
        :return: OrderConfirmation ErrorResponse
        """
        try:
            resp = requests.get('http://example.com', timeout=1)
            resp.raise_for_status()
            print(resp.text)
        except requests.exceptions.Timeout as e:
            print(f"请求超时：{e}")
        except requests.exceptions.RequestException as e:
            print(f"发生了请求异常：{e}")
```

3. 简单性与灵活性之间的平衡

设计模式的一个核心原则是，在简单性与灵活性之间寻找恰当的平衡点。

比如，想要实现一个处理用户注册的 API。在 AOM 中，可以选择将用户注册数据作为单独的参数传递，或者将它们封装在"User"对象（或接口）中，具体如何设计取决于测试的可读性和可维护性。

```python
class UserAPIObject:

    def register_one(self, name: str, email: str, password: str):
        """
        实现一个处理用户注册的API
        :param name:
        :param email:
        :param password:
        :return:
        """
        ...

    def register_two(self, user: dict):
        """
         实现一个处理用户注册的API
        :param user:
        :return:
        """
        name = user.get("name", "")
        email = user.get("email", "")
        password = user.get("password", "")
        ...
```

11.3.2　AOM 使用示例

下面我们通过电商平台的接口示例，演示如何运用 AOM 实施接口自动化测试。

📂 **目录结构**

```
11_chapter/
├──aom
│   ├──__init__.py
│   └──shop_object.py
└──test_shop.py
```

首先，封装电商相关功能的 API 对象。

```python
from random import randint
import json
import requests

BASE_URL = "https://http***.org"

class AuthAPIObject:

    def __init__(self, api_key):
        self.api_key = api_key
```

```python
    def get_token(self, user_id:str) -> str:
        """
        模拟：根据用户ID生成登录token
        :param user_id:
        :return:
        """
        data = {"user_id": user_id, "token": "t" + str(randint(10000, 99999))}
        resp = requests.post(BASE_URL + "/post?key=" + self.api_key, data=data).json()
        return resp["form"]["token"]


class UserAPIObject:

    def __init__(self, token: str):
        self.token = token

    def get_user_data(self, user_id: str):
        """
        模拟：根据用户ID查询用户信息
        :param user_id:
        :return:
        """
        headers = {"token": self.token}
        params = {"user_id": user_id, "name": "tom", "email": "tom@gmail.com", "age": 23}
        resp = requests.post(BASE_URL + "/post", headers=headers, json=params).json()
        return json.loads(resp["data"])


class ProductAPIObject:

    def __init__(self, token: str):
        self.token = token

    def get_product_data(self, product_id: str):
        """
        模拟：根据产品ID查询产品信息
        :param product_id:
        :return:
        """
        headers = {"token": self.token}
        params = {"product_id": product_id, "name": "潮流T恤", "type": "衣服", "price": 69.9}
        resp = requests.post(BASE_URL + "/post", headers=headers, json=params).json()
        return json.loads(resp["data"])
```

📖 **代码说明**

AuthAPIObject 类用于封装与用户认证相关的接口。它通过 api_key 参数接收接口所需的关键 key。get_token()方法用于获取用户登录后生成的 token，我们通过随机数生成一个 token。

UserAPIObject 类用于封装与用户相关的接口。在调用接口之前，需要登录获取 token。get_user_data()方法允许我们通过 user_id 查询和获取用户数据，这里使用预置的数据模拟接口查询返回的数据。

ProductAPIObject 类用于封装与商品相关的接口。在调用接口之前，需要登录获取 token。get_product_data()方法允许我们通过 product_id 查询商品的详细信息，这里同样使用预置的数据模拟接口查询返回的数据。

✍ **使用示例**

下面通过示例演示 API 对象的使用。

```python
import unittest
from aom.shop_object import AuthAPIObject, UserAPIObject, ProductAPIObject

class ShopTest(unittest.TestCase):

    def setUp(self) -> None:
        auth_api = AuthAPIObject("api_key_123")
        self.token = auth_api.get_token("user123")

    def test_user_info(self):
        """
        用户信息查询接口
        """
        user_api = UserAPIObject(self.token)
        user_data = user_api.get_user_data("tom123")
        self.assertEqual(user_data["name"], "tom")

    def test_product_info(self):
        """
        商品信息查询接口
        """
        product_api = ProductAPIObject(self.token)
        product_data = product_api.get_product_data("product123")
        self.assertEqual(product_data["name"], u"潮流 T 恤")

if __name__ == '__main__':
    unittest.main()
```

在编写 API 测试用例过程中，我们只需调用 shop_object.py 文件中封装的类方法，并提供必要的参数即可，无须关注 API 的 URL 地址、请求方法和请求头部（header）信息，以及与核心测试无关的其他参数。当 API 发生变更时，我们只需维护 API Object 层的定义即可。只要对外提供的方法参数保持不变，测试用例层就不需要做任何修改，从而大大简化了接口自动化测试用例的维护工作。

第12章 自动化测试平台化

自动化测试正在朝着平台化方向发展。在性能测试和接口测试这两个领域，平台化的方案已相对成熟，并且得到了广泛应用。然而，对于 Web UI 和 App UI 自动化测试来说，测试用例的编写步骤需要高度的灵活性，而平台化的 UI 交互往往难以满足这一要求。尽管如此，一些开源项目已开始尝试通过平台化来管理和执行测试用例，为提高自动化测试的灵活性和效率提供了可能。

12.1 自动化测试平台化的基本信息

12.1.1 性能测试

目前，性能测试分为前端 Web 性能测试、MWeb 性能测试、后端 API 性能测试和移动端 App 性能测试等。性能测试平台主要以实现后端 API 性能测试为主，集成 JMeter、locust 等开源性能测试工具，通过界面提供性能测试用例的创建、参数配置，以及性能结果的实时展示等。

下面以集成 JMeter 的性能测试平台为例，首先，通过 JMeter 编写测试用例，生成.jmx 性能测试脚本文件。其次，通过平台配置与并发相关的参数，集成 JMeter 性能测试平台，如图 12-1 所示。

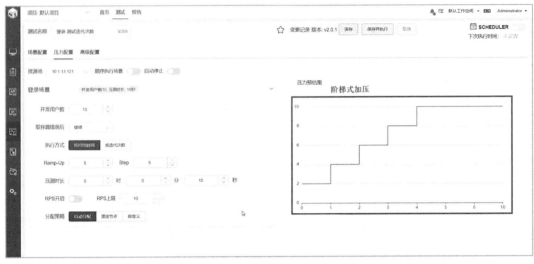

图 12-1　性能测试平台（截图来自开源平台 MeterSphere）

12.1.2　HTTP 接口自动化测试

后端服务的功能主要以 API 的方式提供，应用比较广泛的是使用 HTTP 接口。调用 HTTP 接口有较为固定的格式。

HTTP 接口的基本信息如表 12-1 所示。

表 12-1　HTTP 接口的基本信息

选项	说明	示例
URL	URL（统一资源定位符），即接口的地址	比如，https://www.sam***.com/v1/login
Method	接口的请求方法	比如，GET、POST、PUT和DELETE
Auth	接口的认证	……
Header	接口的请求头	……
参数类型	接口的参数类型	比如，params、form-data、json和xml
参数值	接口的参数值	……
返回值	接口的返回数据，包括状态码、数据格式和数据内容等	……
……	……	……

因为接口的格式较为固定，所以可以很容易地通过界面化实现，并通过接口测试平台提供的接口信息进行配置，如图 12-2 所示。

图 12-2　接口测试平台（截图来自开源平台 MeterSphere）

12.1.3　Web UI 自动化测试

Web UI 自动化测试主要模拟用户在 Web 浏览器上的操作，操作步骤不固定。比如，模拟用户"注册"和"登录"的操作步骤是不同的，甚至同一个网站也会提供多种登录方式，而每种登录方式的操作步骤也不尽相同。

如果将测试步骤进行拆解，那么就变成了针对每个元素的操作。比如，输入"用户名"，输入"密码"，单击"登录"按钮。基于这样的思想，首先，我们可以建立元素库，如图 12-3 所示。

图 12-3　建立元素库（截图来自开源平台 MeterSphere）

其次，将单个元素组装成用例（场景）步骤，管理测试用例，如图 12-4 所示。

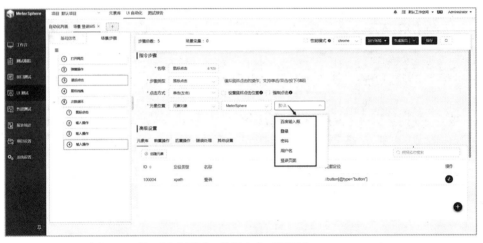

图 12-4　管理测试用例（截图来自开源平台 MeterSphere）

12.1.4　App UI 自动化测试

App UI 自动化测试与 Web UI 自动化测试在本质上是一样的，它们都是模拟用户对应用程序的操作。然而，App UI 自动化测试的环境相对更为复杂，因为它需要接入 Android 或 iOS 设备，或者使用模拟器。此外，需要考虑平台驱动因素，以及在混合应用中上下文的切换。幸运的是，借助像 Appium 这样的跨平台测试工具，这些复杂的问题是可以被有效管理和克服的。

与 Web UI 自动化测试类似，首先，建立控件元素库，如图 12-5 所示。

图 12-5　建立控件元素库（截图来自开源平台 Sonic）

其次，创建测试用例，添加控件元素组装步骤，管理测试用例，如图 12-6 所示。

图 12-6　管理测试用例（截图来自开源平台 Sonic）

12.1.5　自动化测试平台化的优缺点

自动化测试平台化的优缺点如下。

优点：

- 门槛更低：通过 UI 操作就可以完成测试用例的创建、编辑和执行，对于非开发人员更加友好。
- 任务管理：更方便地实现任务管理，如定时任务。
- 监控统计：更方便地实现测试数据管理，如人员和测试用例数量、历史执行数据等。

缺点：

- 场景测试更复杂：在复杂的测试场景中，平台化配置更加烦琐。比如，在接口自动化测试中，不同接口之间存在数据依赖关系，如果想要实现测试数据的依赖，则配置会比较复杂。
- 灵活性较差：在编写测试用例时，我们可以利用封装、继承、条件判断（if 语句）和循环（for 循环）等特性来构建复杂的逻辑结构。然而，这些特性在测试平台上很难完全实现，导致在编写测试用例时，灵活性相对较差。
- 功能扩展不方便：在编程语言中，可以实时设计和封装模块，或者集成第三方库，从而实现功能的快速扩展。然而，采用自动化测试平台是无法实现这种实时性的。我们要考虑如何在 UI 上设计功能的交互，以提高自动化测试平台的可扩展性、易用性和直观性。

12.2 测试框架与测试平台的整合方案

通过前面对自动化测试平台化的介绍，我们认识到，在编写测试用例阶段，采用编程语言编写测试用例可以极大地提高效率和灵活性。而在测试用例编写完成后，在测试用例管理、测试任务管理和测试报告管理等方面，采用平台化的方法则展现出了其无可比拟的优势。那么，是否有一种方案可以同时兼容测试框架与测试平台各自的优势呢？测试框架与测试平台的整合方案如图 12-7 所示。

图 12-7 测试框架与测试平台的整合方案

使用测试框架编写的自动化测试用例最终会生成程序文件（.py）；而基于测试平台创建的自动化测试用例最终会以数据形式保存到数据库中，两者有着本质的区别。

基于 unittest 和 requests 编写的接口测试用例如下。

```python
# test_api_case.py
import unittest
import requests

class TestRequest(unittest.TestCase):

    def setUp(self):
        self.base_url = "http://http***.org"

    def test_put_method(self):
        r = requests.put(self.base_url + '/put', data={'key': 'value'})
        self.assertEqual(r.status_code, 200)

    def test_post_method(self):
        r = requests.post(self.base_url + '/post', data={'key':'value'})
        self.assertEqual(r.status_code, 200)

    def test_get_method(self):
        payload = {'key1': 'value1', 'key2': 'value2'}
        r = requests.get(self.base_url + "/get", params=payload)
        self.assertEqual(r.status_code, 200)

    def test_delete_method(self):
        r = requests.delete(self.base_url + '/delete')
        self.assertEqual(r.status_code, 200)
```

```
if __name__ == '__main__':
    unittest.main()
```

基于测试平台创建的接口测试用例，其数据最终会被保存到数据库中，如图 12-8 所示。

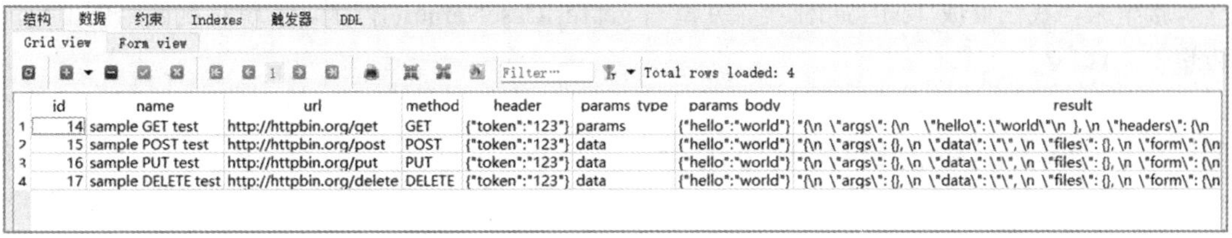

图 12-8　数据库中保存的数据

12.2.1　unittest 解析用例

首先，思考一个问题，unittest 是如何识别程序中的测试用例的？在通过 unittest 编写自动化测试的过程中存在大量"非用例"的方法，比如，setUp()方法、tearDown()方法和自己封装的 login()方法等。unittest 可以精确地识别出测试用例的方法，那么它一定有一套自己的"识别规则"。

unittest 的 TestLoader 类负责根据标准加载测试用例，并将它们包装在测试套件中。

```
...
class TestLoader(object):
    """
    这个类负责根据各种标准加载测试，并以 TestSuite 的形式返回它们
    """
    testMethodPrefix = 'test'
    sortTestMethodsUsing = staticmethod(util.three_way_cmp)
    testNamePatterns = None
    suiteClass = suite.TestSuite
    _top_level_dir = None

    def __init__(self):
        super(TestLoader, self).__init__()
        self.errors = []
        # 追踪通过 load_tests 调用的包
        # 避免无限递归
        self._loading_packages = set()
...
```

📖 代码说明

TestLoader 类位于 unittest 模块的 loader.py 文件中，在类全局变量中，testMethodPrefix 用于定义测试方法（用例）的前缀，这也解释了为什么 unittest 要求测试用例方法名必须以"test"

开头。

整个类的实现比较复杂，想要实现跨文件查找测试用例，则一般会用到 discover()方法。

接下来，我们集成 TestLoader 类，并重写 getTestCaseNames()方法，实现对测试用例信息的收集。

12.2.2 对测试用例的收集与运行

📁 目录结构

```
12_chapter/
├──running          # 实现测试运行器
│   ├──__init__.py
│   ├──loader_extend.py
│   └──runner.py
├──test_dir         # 测试用例目录
│   ├──sub_dir
│   │   ├──__init__.py
│   │   └──test_sub_sample.py
│   ├──__init__.py
│   └──test_sample.py
└──run.py           # 收集并运行测试
```

首先，在 loader_extend.py 文件中创建 MyTestLoader 类。

```
"""
Custom loading unittest.
"""
Import functools
from fnmatch import fnmatchcase
from unittest.loader import TestLoader
class MyTestLoader(TestLoader):
    """
    这个类负责根据各种标准加载测试，并以 TestSuite 的形式返回它们
    """
    testNamePatterns = None
    collectCaseInfo = False    # 控制是否启动测试用例的收集功能
    collectCaseList = []       # 将收集的测试用例信息存储为列表

    def getTestCaseNames(self, testCaseClass):
        返回在 testCaseClass 中找到的方法名称的排序序列
        """

        def shouldIncludeMethod(attrname):
            """
            需包含 Method
            :param attrname:
```

```python
        :return:
        """
        if not attrname.startswith(self.testMethodPrefix):
            return False
        testFunc = getattr(testCaseClass, attrname)
        if not callable(testFunc):
            return False

        fullName = f"""{testCaseClass.__module__}.{testCaseClass.__qualname__}.{attrname}"""

        if self.collectCaseInfo is True:
            case_info = {
                "file": testCaseClass.__module__,
                "class": {
                    "name": testCaseClass.__name__,
                    "doc": testCaseClass.__doc__
                },
                "method": {
                    "name": attrname,
                    "doc": testFunc.__doc__
                }
            }
            self.collectCaseList.append(case_info)

        return self.testNamePatterns is None or \
            any(fnmatchcase(fullName, pattern) for pattern in self.testNamePatterns)

    testFnNames = list(filter(shouldIncludeMethod, dir(testCaseClass)))
    if self.sortTestMethodsUsing:
        testFnNames.sort(key=functools.cmp_to_key(self.sortTestMethodsUsing))
    return testFnNames
```

📖 **代码说明**

创建 **MyTestLoader** 类，使其继承 TestLoader 类，并重写父类中的 getTestCaseNames()方法。

其中，collectCaseInfo 变量用于控制是否启动测试用例的收集功能，而 collectCaseList 变量用于将收集的测试用例信息存储为列表。当 collectCaseInfo 变量被设置为 True 时，开始收集测试用例。

- testCaseClass.__module__：获取测试类的所属模块。
- testCaseClass.__name__：获取测试类的名称。
- testCaseClass.__doc__：获取测试类的描述。
- attrname：测试方法名称。
- testFunc.__doc__：测试方法的描述。

最终，将上面的数据以 dict 格式保存到 collectCaseList 列表中。

接下来，在 runner.py 文件中实现 TestMain 类。

```python
import unittest
import json
from running.loader_extend import MyTestLoader

class TestMain:
    """
    TestMain 测试类
    1. 收集测试用例信息并返回到列表
    2. 根据列表运行测试用例
    """
    TestSuits = []

    def __init__(self, path: str = None):
        """
        运行测试用例
        :param path:
        """
        self.path = path
        if path is None:
            raise FileNotFoundError("Specify a file path")

        self.TestSuits = MyTestLoader().discover(start_dir=self.path)

    @staticmethod
    def run(suits):
        """
        运行测试用例
        """
        runner = unittest.TextTestRunner()
        runner.run(suits)

    @staticmethod
    def collect_cases(is_json: bool = False):
        """
        返回收集的测试用例信息
        MyTestLoader.collectCaseInfo = True
        :param is_json: Return JSON format
        """
        cases = MyTestLoader.collectCaseList

        if is_json is True:
            return json.dumps(cases, indent=2, ensure_ascii=False)

        return cases

    def _load_testsuite(self):
```

```python
        """
        加载测试套件并转换为映射
        """
        mapping = {}

        for suits in self.TestSuits:
            for cases in suits:
                if isinstance(cases, unittest.suite.TestSuite) is False:
                    continue

                for case in cases:
                    file_name = case.__module__
                    class_name = case.__class__.__name__

                    key = f"{file_name}.{class_name}"
                    if mapping.get(key, None) is None:
                        mapping[key] = []

                    mapping[key].append(case)

        return mapping

    def run_cases(self, data: list):
        """
        运行测试用例列表
        :param data: test case list
        """
        if isinstance(data, list) is False:
            raise TypeError("Use cases must be lists.")

        if len(data) == 0:
            raise ValueError("There are no use cases to execute")

        suit = unittest.TestSuite()

        case_mapping = self._load_testsuite()
        print(case_mapping)
        for d in data:
            d_file = d.get("file", None)
            d_class = d.get("class").get("name", None)
            d_method = d.get("method").get("name", None)
            if (d_file is None) or (d_class is None) or (d_method is None):
                raise NameError("""Use case format error""")

            cases = case_mapping.get(f"{d_file}.{d_class}", None)
            if cases is None:
                continue

            for case in cases:
                method_name = str(case).split(" ")[0]
```

```
                if method_name == d_method:
                    suit.addTest(case)

        self.run(suit)
```

📖 **代码说明**

定义 TestMain 类，path 参数用于指定测试目录。调用 MyTestLoader 类的 discover()方法，可以收集 path 目录中的测试用例，将收集的测试用例保存到 TestSuits 变量中。

run()方法通过调用 unittest.TextTestRunner 类，运行测试套件中的测试用例。

collect_cases()方法返回 MyTestLoader.collectCaseList 中收集的测试用例。如果 is_json 变量为 True，则将结果转换为 JSON 格式。

_load_testsuite()方法主要对 TestSuits 变量中收集的测试用例进行处理：找出测试用例的文件名和类名，并组成字典 key，添加测试用例到 value 列表中，格式如下。

```
{
  'sub_dir.test_sub_case.TestSubDirCase':   # 目录文件名
  [  # 测试用例对象列表
    <sub_dir.test_sub_case.TestSubDirCase testMethod=test_sub_case>
  ],
  'test_sample.TestSample':
  [
    <test_sample.TestSample testMethod=test_case_01>,
    <test_sample.TestSample testMethod=test_case_02>
  ]
}
```

run_cases()方法用于运行测试用例，data 为收集的测试用例数据，处理过程大致如下。

将调用_load_testsuite()方法返回的测试用例与 data 中的测试用例进行对比，如果文件名、类名和方法名完全相同，则通过 TestSuite 类提供的 addTest()方法添加测试用例至测试列表。最后，调用 run()方法运行测试套件中的测试用例。

🔖 **使用示例**

验证 running 模块中实现的功能。首先，在 test_dir 目录中准备一些测试用例。

下面是根目录测试用例文件中的内容（在 test_dir/test_sample.py 文件中）。

```
import unittest

class TestSample(unittest.TestCase):
    """简单的测试类"""

    def test_case_01(self):
        """测试用例 01"""
```

```python
        self.assertEqual(1 + 1, 2)

    def test_case_02(self):
        """测试用例 02"""
        self.assertEqual(2 + 2, 4)

if __name__ == '__main__':
    unittest.main()
```

下面是子目录测试用例文件中的内容（在 test_dir/sub_dir/test_sub_sample.py 文件中）。

```python
import unittest

class TestSubDirCase(unittest.TestCase):
    """子目录测试类"""

    def test_sub_case(self):
        """测试子目录中的测试用例"""
        self.assertEqual(4 + 4, 8)

if __name__ == '__main__':
    unittest.main()
```

创建 run.py 文件，用于收集测试用例。

```python
from running.loader_extend import MyTestLoader
from running.runner import TestMain

if __name__ == '__main__':
    # 启动测试用例收集功能
    MyTestLoader.collectCaseInfo = True
    # 指定运行测试用例目录
    test_main = TestMain(path="./test_dir/")
    # 收集测试用例
    case_info = test_main.collect_cases(is_json=True)
    print(case_info)
```

📖 **代码说明**

设置 MyTestLoader 类下面的 collectCaseInfo 变量为 True，启动收集测试用例功能。

调用 TestMain 类，通过 path 参数设置收集测试用例的目录，调用 collect_cases()方法收集测试用例。如果 is_json 参数为 True，则将收集的测试用例数据转换为 JSON 格式。

⏳ **运行结果**

```
> python run.py
[
  {
    "file": "sub_dir.test_sub_case",
    "class": {
```

```
      "name": "TestSubDirCase",
      "doc": "子目录测试类"
    },
    "method": {
      "name": "test_sub_case",
      "doc": "test sub dir case"
    }
  },
  {
    "file": "test_sample",
    "class": {
      "name": "TestSample",
      "doc": "简单的测试类"
    },
    "method": {
      "name": "test_case_01",
      "doc": "test case 01"
    }
  },
  {
    "file": "test_sample",
    "class": {
      "name": "TestSample",
      "doc": "简单的测试类"
    },
    "method": {
      "name": "test_case_02",
      "doc": "test case 02"
    }
  }
]
.
----------------------------------------------------------------------
Ran 1 test in 0.000s

OK
```

通过运行结果可以看出，用例的"目录和文件名""类名和类注释""方法名和方法注释"都已经收集到了。我们可以将测试用例数据保存到数据库表中，进而在测试平台上展示。

修改 run.py 文件，通过配置测试用例信息来运行测试用例。

```
from running.loader_extend import MyTestLoader
from running.runner import TestMain

if __name__ == '__main__':
    # 收集的测试用例列表
    cases = [
      {
        "file": "sub_dir.test_sub_case",
```

```
      "class": {
        "name": "TestSubDirCase",
        "doc": "子目录测试类"
      },
      "method": {
        "name": "test_sub_case",
        "doc": "test sub dir case"
      }
    }
]
# 指定运行测试用例的目录
test_main = TestMain(path="./test_dir")
# 运行收集的测试用例
test_main.run_cases(cases)
```

> **代码说明**

参考收集的测试用例格式，在 cases 中定义测试用例信息，调用 TestMain 类。path 参数指定了运行测试用例的目录，调用 run_cases()方法运行 cases 中的测试用例。

> **运行结果**

```
> python run.py
.
----------------------------------------------------------------------
Ran 1 test in 0.000s

OK
```

通过运行结果可以看到，指定的测试用例被运行了。在测试平台上，我们可以将收集的测试用例信息保存到数据库中，当需要运行测试用例时，再将其从数据库中读取出来，调用 TestMain 类中的 run_cases()方法来运行。

至此，我们已经实现了测试框架与测试平台的整合方案的核心思路。

12.3 SeldomQA 项目

在基于 unittest 实现了对测试用例数据的获取后，就具备了平台化的基础。通过使用 Web 开发技术，可以将测试框架编写的测试用例解析到平台上进行展示和运行。由于本书的重点是介绍测试框架的设计，而 Web 开发技术展开来讲会是一个庞大的技术栈。基于这个核心实现，笔者开源了测试框架和平台相关项目。

12.3.1 Seldom 框架

Seldom 框架是基于 unittest 的全功能自动化测试框架,它支持 Web、App 和 API 等类型的测试。通过 Seldom 框架,我们可以方便地编写自动化测试用例。

用 pip 命令安装 Seldom 框架。

```
> pip install seldom
```

1. Web 测试

Seldom 框架集成了 Selenium 框架,并对 Selenium 的 API 进行了二次封装,同时为 Web 测试专门设计了断言方法。这使得编写 Web 自动化测试用例变得更加简单。此外,Seldom 框架支持全局浏览器的启动和关闭,以及元素的智能等待机制,这些都有助于提高 Web 自动化测试用例的运行速度和稳定性。

```python
import seldom

class WebTest(seldom.TestCase):

    def test_bing_search(self):
        """
        bing web search
        """
        self.open("https://cn.b***.com")
        self.type(id_="sb_form_q", text="seldom", enter=True)
        self.assertTitle("seldom - 搜索")

if __name__ == '__main__':
    seldom.main(browser="chrome")
```

⌛ 运行结果

运行 Web 自动化测试用例。

```
> python test_web.py
```

```
                      _      _
    _____     / /____/ /___  ____ ___
   / ___/ _ \   / / __  / __ \/ __ `__ \
  (__  )  __/_ / / /_/ / /_/ / / / / / /
 /____/\___/(_)_/\__,_/\____/_/ /_/ /_/  v3.7.0
------------------------------------------
                @itest.info

XTestRunner Running tests...
----------------------------------------------------------------------
2024-05-14 00:02:58 | INFO     | webdriver.py | 🔲 https://cn.b***.com
2024-05-14 00:02:58 | INFO     | webdriver.py | ☑ Find 1 element: id=sb_form_q ->
input 'seldom'.
```

```
2024-05-14 00:02:59 | INFO    | case.py   | 🚀 assertTitle -> seldom - 搜索.
Generating HTML reports...
.12024-05-14 00:02:59 | SUCCESS | runner.py | generated html file:
file:///D:\github\test-framework-design\12_chapter\seldom_demo\reports\2024_05_
14_00_02_55_result.html
2024-05-14 00:02:59 | SUCCESS | runner.py | generated log file:
file:///D:\github\test-framework-design\12_chapter\seldom_demo\reports\seldom_l
og.log
```

2. HTTP 接口测试

Seldom 框架集成了 requests 库,并对 requests 的 API 进行了二次封装,以增强其功能。此外,Seldom 框架为 HTTP 接口测试提供了一套强大的断言方法和详尽的运行日志功能,提升了接口测试的效率和准确性。

```
import seldom


class APITest(seldom.TestCase):

    def test_post_method(self):
        self.post('/post', data={'key':'value'})
        self.assertStatusCode(200)

    def test_get_method(self):
        payload = {'key1': 'value1', 'key2': 'value2'}
        self.get("/get", params=payload)
        self.assertStatusCode(200)

if __name__ == '__main__':
    seldom.main(base_url="http://http***.org")
```

⏳ 运行结果

运行 HTTP 自动化测试用例。

```
> python test_api.py
              __     __
   _____  / /____/ /___  ____ ___
  / ___/ _ \/ / __  / __ \/ __ `__ \
 (__  )  __/ / /_/ / /_/ / / / / / /
/____/\___/_/\__,_/\____/_/ /_/ /_/  v3.7.0
-----------------------------------------
                 @itest.info

XTestRunner Running tests...

----------------------------------------------------------------------
2024-05-14 00:18:38 | INFO     | request.py | -------------- Request
-----------------[🚀]
2024-05-14 00:18:38 | INFO     | request.py | [method]: GET      [url]:
http://http***.org/get
2024-05-14 00:18:38 | DEBUG    | request.py | [params]:
```

```
{
  "key1": "value1",
  "key2": "value2"
}
2024-05-14 00:18:39 | INFO    | request.py | -------------- Response --------------[🚀]
2024-05-14 00:18:39 | INFO    | request.py | successful with status 200
2024-05-14 00:18:39 | DEBUG   | request.py | [type]: json        [time]: 0.459631
2024-05-14 00:18:39 | DEBUG   | request.py | [response]:
{
  "args": {
    "key1": "value1",
    "key2": "value2"
  },
  "headers": {
    "Accept": "*/*",
    "Accept-Encoding": "gzip, deflate",
    "Host": "http***.org",
    "User-Agent": "python-requests/2.31.0",
    "X-Amzn-Trace-Id": "Root=1-66423d5f-65796bb44f61c6245c46b326"
  },
  "origin": "14.155.62.206",
  "url": "http://http***.org/get?key1=value1&key2=value2"
}
2024-05-14 00:18:39 | INFO    | case.py | 🎯 assertStatusCode -> 200.
.12024-05-14 00:18:39 | INFO    | request.py | -------------- Request --------------[🔧]
2024-05-14 00:18:39 | INFO    | request.py | [method]: POST       [url]: http://http***.org/post
2024-05-14 00:18:39 | DEBUG   | request.py | [data]:
{
  "key": "value"
}
2024-05-14 00:18:39 | INFO    | request.py | -------------- Response --------------[🚀]
2024-05-14 00:18:39 | INFO    | request.py | successful with status 200
2024-05-14 00:18:39 | DEBUG   | request.py | [type]: json        [time]: 0.449655
2024-05-14 00:18:39 | DEBUG   | request.py | [response]:
{
  "args": {},
  "data": "",
  "files": {},
  "form": {
    "key": "value"
  },
  "headers": {
    "Accept": "*/*",
    "Accept-Encoding": "gzip, deflate",
    "Content-Length": "9",
    "Content-Type": "application/x-www-form-urlencoded",
    "Host": "httpbin.org",
```

```
    "User-Agent": "python-requests/2.31.0",
    "X-Amzn-Trace-Id": "Root=1-66423d60-2eaade1a0821523775412550"
  },
  "json": null,
  "origin": "14.155.62.206",
  "url": "http://http***.org/post"
}
2024-05-14 00:18:39 | INFO     | case.py   | 🔟 assertStatusCode -> 200.
Generating HTML reports...
.22024-05-14 00:18:39 | SUCCESS  | runner.py | generated html file:
file:///D:\github\test-framework-design\12_chapter\seldom_demo\reports\2024_05_
14_00_18_38_result.html
2024-05-14 00:18:39 | SUCCESS  | runner.py | generated log file:
file:///D:\github\test-framework-design\12_chapter\seldom_demo\reports\seldom_l
og.log
```

3. App 测试

Seldom 框架集成了 Appium。Seldom 框架提供了一些实用的类，比如，Switch 类用于切换上下文，Action 类提供了触摸和滑动等操作，FindByText 类允许基于文本来查找元素，以及 KeyEvent 类允许基于键盘输入字符串。

在使用 Seldom 框架进行 App 测试时，需要依赖 Appium 的运行环境,包括启动 appium-server 以监听特定端口，以及准备所需的手机设备来运行测试等。

```python
import seldom
from appium.options.android import UiAutomator2Options
from seldom.appium_lab.keyboard import KeyEvent

class TestBingApp(seldom.TestCase):

    def start(self):
        self.ke = KeyEvent(self.driver)

    def test_bing_search(self):
        """
        test bing App search
        """
        self.sleep(2)
        self.click(id_="com.microsoft.bing:id/sa_hp_header_search_box")
        self.type(id_="com.microsoft.bing:id/sapphire_search_header_input",
text="seldomQA")
        self.ke.press_key("ENTER")
        self.sleep(1)
        elem = self.get_elements(xpath='//android.widget.TextView')
        self.assertIn("seldom", elem[0].text.lower())


if __name__ == '__main__':
    capabilities = {
        'deviceName': 'ELS-AN00',
```

```python
        'automationName': 'UiAutomator2',
        'platformName': 'Android',
        'appPackage': 'com.microsoft.bing',
        'appActivity': 'com.microsoft.sapphire.app.main.MainSapphireActivity',
        'noReset': True,
    }
    options = UiAutomator2Options().load_capabilities(capabilities)
    seldom.main(app_server="http://127.0.0.1:4723", app_info=options)
```

3 运行结果

运行 App 自动化测试用例。

```
> python test_app.py
      _____ ___ _____ _____  _____
     / ____// /  / __ \/   |/  / ___/
    / (___ / _ \/ / / / / /| |/ /__  \
    \___ \/  __/ / / / / /_| / / / __/
   /____/\___/_/_/_/\__,_/_/  /_/  v3.7.0
   --------------------------------------
                    @itest.info

XTestRunner Running tests...
----------------------------------------------------------------
2024-05-14 00:33:33 | INFO    | case.py     | 💤 sleep: 2s.
2024-05-14 00:33:36 | INFO    | webdrivor.py| ☑ Find 1 element:
id=com.microsoft.bing:id/sa_hp_header_search_box -> click.
2024-05-14 00:33:38 | INFO    | webdriver.py| ☑ Find 1 element:
id=com.microsoft.bing:id/sapphire_search_header_input -> input 'seldomQA'.
2024-05-14 00:33:38 | INFO    | keyboard.py | press key "ENTER"
2024-05-14 00:33:39 | INFO    | case.py     | 💤 sleep: 1s.
2024-05-14 00:33:42 | INFO    | webdriver.py| ☑ Find 33 element:
xpath=//android.widget.TextView .
Generating HTML reports...
.12024-05-14 00:33:42 | SUCCESS | runner.py | generated html file:
file:///D:\github\test-framework-design\12_chapter\seldom_demo\reports\2024_05_
14_00_33_27_result.html
2024-05-14 00:33:42 | SUCCESS | runner.py | generated log file:
file:///D:\github\test-framework-design\12_chapter\seldom_demo\reports\seldom_l
og.log
```

12.3.2　seldom-platform

seldom-platform 是基于 Seldom 框架的平台化项目，它可以方便地接入并使用 Seldom 框架编写的自动化测试用例。

1. 启动 seldom-platform

seldom-platform 由前端和后端两部分组成，前端主要基于 Vue.js 和 Naive-UI 等技术实现，而后端则主要采用 Django 框架和 django-ninja 库进行开发。

在运行 seldom-platform 之前，需要克隆项目到本地。

```
> git clone https://git***.com/SeldomQA/seldom-platform    # 克隆项目
> cd seldom-platform    # 进入项目目录
```

启动 Redis。因为后端服务依赖 Redis，所以需要安装 Redis 并启动。

```
> redis-server
```

启动后端服务，默认占用 8000 端口。

```
> cd backend    # 进入后端目录
> pip install -r requirements.txt    # 安装相关依赖
> python manage.py runserver    # 启动后端服务
```

启动前端服务，默认占用 5173 端口，前提是先安装 Node.js。

```
> cd frontendv3    # 进入前端目录
> pnpm install    # 安装相关依赖
> pnpm dev    # 启动前端服务
```

在前后端项目都正常启动后，通过浏览器访问 http://localhost:5173，默认登录账号和密码分别是"guest"和"guest123456"。

2. 使用平台功能

首先，通过 Seldom 框架编写自动化测试项目，并使用 Git 的代码托管平台（如 Gitee、GitHub 和 GitLab）进行管理。然后，在 seldom-platform 上接入 Git 项目。

准备一个 Seldom API 自动化测试项目。

进入菜单"项目管理"，打开"新建项目"对话框，如图 12-9 所示。

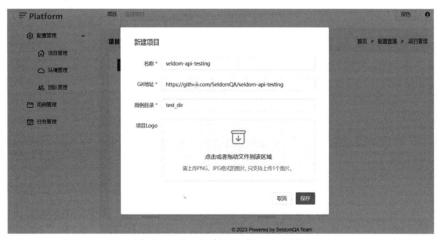

图 12-9　打开"新建项目"对话框

参数说明。

- 名称：项目名称。
- Git 地址：Gitee、GitHub 或 GitLab 等 Git 仓库地址。
- 用例目录：指测试项目中的测试用例目录。比如，seldom-api-testing 项目中的 test_dir/ 目录，就是测试用例目录。

单击左侧的"用例管理"选项，在右侧页面选择新建的项目，单击"同步"按钮同步用例，如图 12-10 所示。

图 12-10　同步用例

同步用例的过程共分为四步。

第 1 步，拉取代码。如果是首次执行项目，则克隆项目（用 git clone 命令）到本地（平台部署的主机）。如果已经克隆过项目，则拉取（用 git pull 命令）最新的代码到本地。

第 2 步，同步用例。使用 Seldom 框架提供的 API 解析项目中的测试用例，并将其保存到数据库临时表。

第 3 步，查找结果。将临时表中的测试用例与正式表中的测试用例进行对比，显示新增或删除的测试用例。

第 4 步，合并结果。将临时表中的测试用例更新到正式表中。

注意：平台同步项目需要在 Git 环境中进行，并且必须具备对自动化测试项目进行克隆和拉取的操作权限。

用例管理如图 12-11 所示，从图中可以查看同步的 Seldom 自动化测试项目中的测试用例。

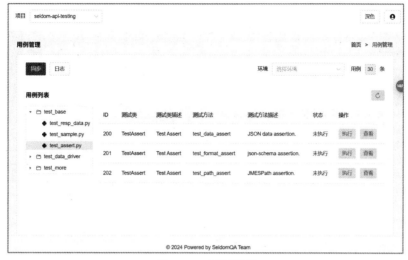

图 12-11　用例管理

平台解析了 Seldom 自动化测试项目中的测试目录、测试类和测试方法，以及类和方法的注释信息，如图 12-12 所示。

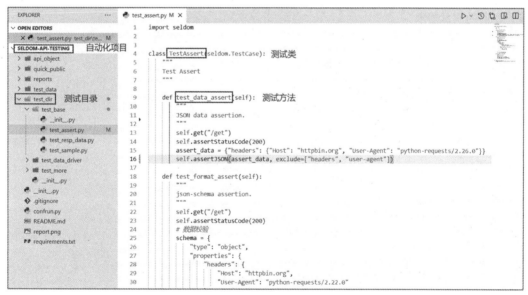

图 12-12　自动化代码

单击"执行"按钮运行测试用例，在测试用例被运行后，单击"查看"按钮查看运行结果，如图 12-13 所示。

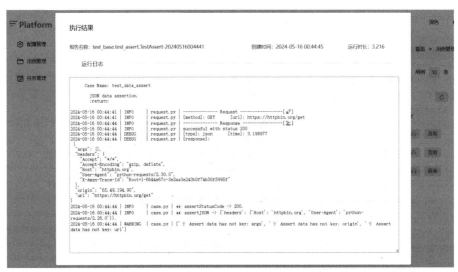

图 12-13 运行结果

至此，seldom-platform 最核心的功能已介绍完毕。除此之外，seldom-platform 集成了任务管理和测试报告等实用功能。在此基础上，我们可以进一步开发和扩展更多定制化的功能，以满足不同的测试需求。

第 13 章

自动化测试的 AI 探索

近几年,自动化测试正朝着 AI 方向发展,希望通过 AI 技术解决自动化测试过程中的一些痛点。比如,自动化测试的编写难度、效率,以及用例的稳定性等。本章我们一起探讨 AI 技术在自动化测试领域的最新进展,并了解相关的平台、项目和工具。

13.1 集成 AI 技术的自动化测试平台

目前国外一些公司正在尝试通过 AI 技术为自动化测试赋能,已经提供了成熟产品的有 Applitools、Testim、mabl 和 Eggplant 等。这些产品无疑都是通过平台化的方式提供功能和服务的。因为 AI 是需要不断学习和进化的技术,同时需要算力的支撑,所以只有提供平台化才能满足这样的要求。

13.1.1 基于视觉 AI 技术的自动化检测

Applitools 是一家专注于视觉 AI 技术的提供自动化测试和监控解决方案的公司,其核心产品主要集中在视觉回归测试和 UI 测试领域。

Applitools Eyes 是其旗舰产品,是一个基于 AI 技术的自动化测试工具,专注于检测和修复 UI 错误。它使用视觉 AI 技术识别界面的变化,并能够跨浏览器和设备进行比较,确保 UI 的一致性和准确性。

1. 注册 Applitools 平台账号

想要使用 Applitools Eyes 的功能,就要注册 Applitools 平台账号,Applitools 平台要求必须使用企业邮箱。

Applitools Eyes 提供了对 Web、Mobile 和 Desktop 等平台的测试，并支持主流的开源测试库。

- Web 平台支持：Selenium、Cypress、Playwright 和 Puppeteer 等。
- Mobile 平台支持：Appium、Expresso 和 XCUI 等。
- Desktop 平台支持：CodedUI。

此外，Applitools Eyes 支持对 Storybook、图片和 PDF 文件等的测试，Applitools 提供的功能如图 13-1 所示。

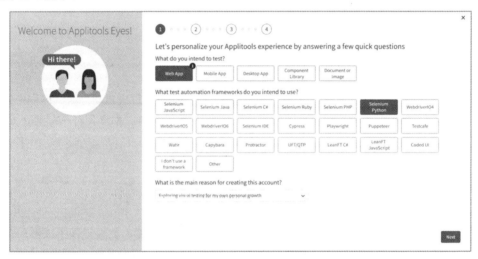

图 13-1　Applitools 提供的功能

在使用 Applitools Eyes 之前，需要使用企业邮箱注册一个账号。根据当前官方政策，新用户可以享受 15 天的免费试用期。注册账号并登录后，可进入测试管理页面。在页面右上角单击账号名称，在弹出的下拉列表中选择"My API key"选项，如图 13-2 所示，将弹出一个窗口，显示"My API key"字符串。

图 13-2　选择"My API key"选项

2. 编写自动化测试脚本

Applitools Eyes 扩展了对主流测试库的支持,并提供了一系列用于视觉测试的 API,如 eyes-selenium 和 eyes-playwright。

用 pip 命令安装 eyes-selenium。

```
> pip install eyes-selenium
```

◆ 使用示例

创建 eyes_selenium_demo.py 文件,编写基于 eyes-selenium 的自动化测试脚本。

```python
from applitools.selenium import Eyes, Target, Configuration
from selenium import webdriver
from selenium.webdriver.common.by import By

# 初始化 Selenium WebDriver
driver = webdriver.Chrome()

# 初始化 Applitools Eyes
eyes = Eyes()
eyes.api_key = '{api key}'

try:
    # 创建一个新的测试实例配置
    config = Configuration()
    config.app_name = 'Applitools Hello World Demo'
    config.test_name = 'Hello World Test with Links and Button'

    # 开始视觉测试
    with eyes.open(
            driver, app_name="Hello World App",
            test_name="Hello World Test",
            viewport_size={'width': 800, 'height': 600}):

        # 访问目标页面
        driver.get("https://applit***.com/helloworld")

        # 检测主页面
        eyes.check("Main Page", Target.window())

        # 单击第一个 diff 链接
        driver.find_element(By.CSS_SELECTOR, 'a[href="?diff1"]').click()
        eyes.check("Diff1 Page", Target.window())

        # 返回主页面
        driver.back()
```

```
        # 单击第二个 diff 链接
        driver.find_element(By.CSS_SELECTOR, 'a[href="?diff2"]').click()
        eyes.check("Diff2 Page", Target.window())

        # 返回主页面
        driver.back()

        # 单击按钮
        driver.find_element(By.TAG_NAME, 'button').click()

        # 检测按钮被单击后的页面
        eyes.check("After Button Click", Target.window())

        # 结束视觉测试并关闭浏览器
        eyes.close()
finally:
    driver.quit()
    eyes.abort()
```

📖 **代码说明**

首先,实例化 Eyes 类并将其赋值给 eyes 变量,通过 eyes.api_key 指定 Applitools Eyes 平台生成的 API key,然后,调用 eyes.open()方法开始视觉测试。driver 用于指定 Selenium 浏览器驱动,app_name 用于指定被测试应用的名称,test_name 用于指定当前测试用例的名称,viewport_size 用于设置屏幕的尺寸。

此外,可以调用 Configuration 类创建全局的测试实例配置。app_name 用于指定被测试应用的名称,test_name 用于指定测试用例的名称。这个配置是全局的,多个测试可以共享相同的配置,而不必在每个测试中重复设置相同的参数。

最核心的是 eyes.check()方法,主要用于进行视觉检测,其中,Target.window()表示检测整个窗口。

3. 运行自动化测试脚本

运行 eyes_selenium_demo.py 自动化测试脚本。

```
> python eyes_selenium_demo.py
```

整个运行过程与单纯的 Selenium 脚本相比,速度略慢,因为包含生成界面快照,以及上传的过程。

(1)第一次运行。

首先,加载页面并调用 eyes.check("xx", Target.window())来捕捉整个页面快照。然后,将快照上传到 Applitools 云,并存储为基准图像。

在第一次运行时,该快照将作为参照基准,无论结果如何,都不会出现视觉检测失败的情况。

(2)第二次运行(以及后续运行)。

再次加载页面,捕捉当前页面的快照,将新快照上传到云,并与之前的基准图像进行智能对比。Applitools Eyes 提供的 Visual AI(视觉 AI)会进行系统检测和差异分析,从而发现快照之间的变化。两次运行的结果如图 13-3 所示,页面中间的随机数导致第二次运行失败。

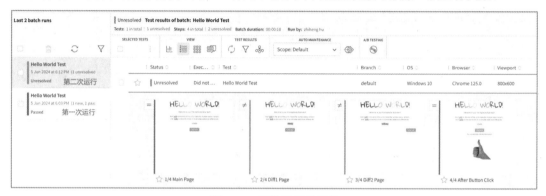

图 13-3　两次运行的结果

(3)差异处理。

开发者在查看报告时,应判断这些变化是预期的还是错误的。如果变化是预期的(如布局调整或价格修改),则应更新基准图像。如果变化是错误的,则开发者应立即修复问题。

对比失败区域如图 13-4 所示,失败的原因是页面中间有一串随机数。当我们确定可以忽略检测随机数时,可以点击失败的图片,通过"IGNORE"工具选择忽略的区域并保存,这样在下次执行时,将忽略对这个区域的检测。

图 13-4　对比失败区域

此外,在对比结果中,第二个 diff 页中的"Check me!"按钮的位置发生了改变,视觉检测会忽略这个改变。这也充分证明了 Applitools Eyes 的能力，在它大量的训练过程中，按钮的位置发生了改变并不是错误，因此它在检测时会忽略。

（4）第三次运行（以及后续运行）。

经过差异处理，再次加载页面，捕捉当前页面的快照。随后，将新快照上传到云，与之前的基准图像进行对比。由于已经添加了忽略区域，所以 Visual AI 会跳过忽略区域。第三次运行是通过的，第三次运行的结果如图 13-5 所示。

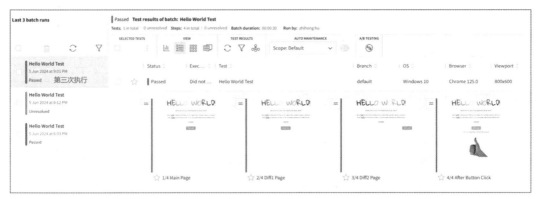

图 13-5　第三次运行的结果

4. Applitools Eyes 的工作过程

与传统的图片对比工具相比，Applitools Eyes 确实具备 AI 能力，那么 Applitools Eyes 是如何工作的？Applitools Eyes 的工作过程如图 13-6 所示。

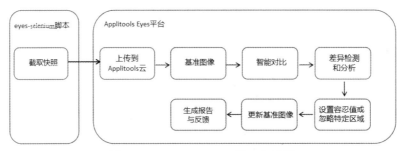

图 13-6　Applitools Eyes 的工作过程

主要步骤如下。

（1）截取快照（Snapshots）。

当调用 eyes.check(...) 方法时，Applitools Eyes 会通过驱动（如 Selenium WebDriver）截取当前页面或视图的屏幕快照。这些快照会捕捉整个页面、特定部分或者具体元素，取决于你配置的检测目标（比如，整个窗口或某个特定元素）。

(2) 上传到 Applitools 云。

截取的快照将被上传到 Applitools 云。在云端，Applitools 会对这些图像进行处理和优化，以便后续的对比和分析。

(3) 基准图像（Baseline Images）。

在第一次运行测试时，Applitools Eyes 会将这些快照存储为基准图像。后续的每次测试都将生成新的快照，并将这些快照与基准图像进行对比。

(4) 智能对比（Smart Comparison）。

Applitools 使用了一种被称为 Visual AI 的技术来进行智能对比。它采用了计算机视觉和机器学习算法，能够识别和处理图像中的元素和细节。除了简单的像素对比，视觉 AI 可以了解不同上下文中的 UI 元素，识别布局、样式和内容的变化。

(5) 差异检测和分析（Difference Detection and Analysis）。

通过智能对比，Applitools Eyes 可以识别图像中的差异。这些差异可以是位置变化、样式变化或内容变化等。AI 系统会标记并分类这些差异，将其区分为"重大问题"、"微小变化"或"无关紧要"。

(6) 设置容忍值或忽略特定区域。

在某些情况下，用户可以设置容忍值或忽略某些特定区域，以避免不必要的测试失败。比如，动态内容（广告、日期和时间等）可以被标记为忽略区域。

(7) 更新基准图像。

当合法的 UI 变化被接受时，可以将这些测试结果更新为新的基准图像。如此一来，下一次的测试将基于最新的 UI 状态进行对比。

(8) 生成报告与反馈。

Applitools Eyes 会生成详细的报告，列出所有发现的差异，并提供视觉反馈。当然，这些报告可以集成到 DevOps 流水线中，帮助开发和测试团队快速识别和修复 UI 问题。

5. 关于 Applitools 的 AI 能力

Applitools 的 AI 经过了上亿张图片的训练，它不依赖于像素点的直接对比来分析图片（因为这会产生大量负面的正例），而是通过模拟人眼去识别图像之间的错误。人眼会忽略的错误，它也会忽略；而人眼容易发现的错误，它也能识别出来。

目前 Applitools 的识别准确率高达 99.9999%，这意味着用它进行一百万次测试，只能找出一个负面的正例。由此可以看出，Applitools 并没有偏离真正的测试，它将误报率看得非常重。虽然说"人眼算法"听起来有点神奇，但不否认它的图像识别技术确实有独到之处。

总体来说，我们在进行 UI 自动化测试时，要先区分测试的重点是"功能的可用性"还是"视觉检测"，或者二者兼有。然后把需要"视觉检测"的页面交由 Visual AI 来完成是一个不错的选择。

13.1.2　基于 AI 的自动化测试运行

mabl 是一个综合性的自动化测试平台，它的目标是简化和加速软件测试过程。它利用人工智能和机器学习技术，提供智能化、自动化的测试解决方案。

与传统基于页面属性（id、name 和 XPath 等）的自动化测试相比，它的主要优势在于它利用 AI 能力运行自动化测试，从而大大地提高了自动化测试用例运行的稳定性。

1. mabl 平台注册

在使用 mabl 平台之前，需要在 mabl 平台上注册账号。注册时 mabl 平台要求必须使用企业邮箱。

mabl 平台支持的测试类型如图 13-7 所示，包括 Browser test、Mobile test、API test 和 Performance test 等。

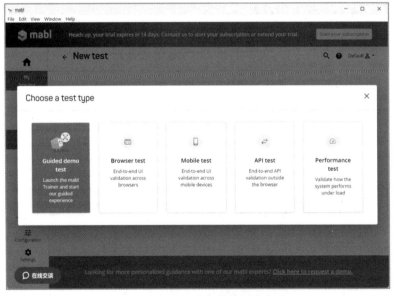

图 13-7　mabl 平台支持的测试类型

mabl 平台以客户端的方式为用户提供服务，为了正常地创建和运行测试，用户在登录 mabl 平台之前需要下载并安装客户端。mabl 平台要求必须安装 Chrome 浏览器，以保证服务能够正常运行。mabl 平台的下载页面如图 13-8 所示。

图 13-8　mabl 平台的下载页面

2．自动化测试用例

对于 Web UI 自动测试，mabl 平台提供了测试用例的录制功能，不需要用户手动编写自动化测试用例或脚本。

首先，通过账号登录 mabl 客户端。系统会提示创建一个低代码测试，并提供了两种测试类型，如图 13-9 所示。此时单击"Browser demo test"选项，mabl 平台会启动 mabl Trainer，同时打开 Chrome 浏览器。

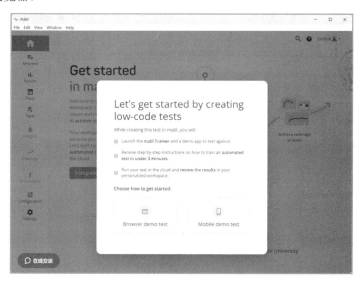

图 13-9　两种测试类型

mabl Trainer 录制测试用例如图 13-10 所示。

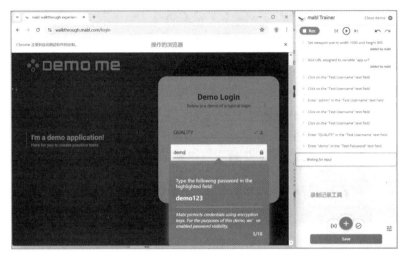

图 13-10　mabl Trainer 录制测试用例

默认录制的是 mabl 平台的官方 demo，根据提示，一步步完成一个网站的登录和退出，包括对测试用例的断言。mabl Trainer 会记录整个操作过程，并提供回放功能。单击 mabl Trainer 页面的"Save"按钮，保存测试用例。

在 mabl 客户端的 tests 管理中可以找到录制好的自动化测试用例，如图 13-11 所示。

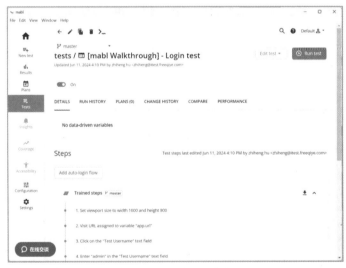

图 13-11　查看测试用例

mabl 平台提供了编辑（Edit test）和运行（Run test）测试用例的功能。

- Edit test：随着被测试 App 的功能变动，录制的测试用例可能会遇到无法运行的情况，此时可以利用编辑功能重新录制和修改测试用例。

- Run test：运行测试分为 Local run（本地运行）和 Cloud run（云端运行）两种。顾名思义，本地运行就是在本地启动浏览器运行测试，云端运行则是利用 mabl 平台提供的远程环境运行测试。

接下来，运行自动化测试用例。单击"Run test"按钮，在弹出的侧边栏中选择运行方式，如图 13-12 所示。

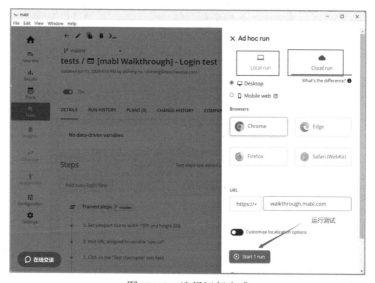

图 13-12　选择运行方式

测试用例运行完成后，单击 mabl 客户端中的"Results"选项，查看运行结果，如图 13-13 所示。

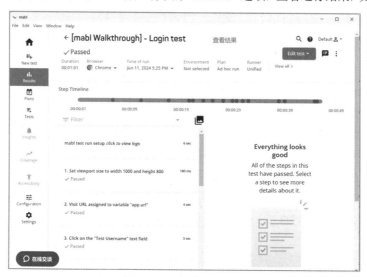

图 13-13　查看运行结果

3. 验证 AI 元素定位的能力

在使用 mabl 平台后，我们认为它所提供的自动化录制和回放功能并不新颖，因为类似的工具早在十几年前就已经出现了，如 Selenium IDE。

接下来，我们将探索 mabl 平台在 AI 方面的能力。为了便于修改页面的元素和样式，笔者在本地启动一个了 Web 应用，并采用了第 12 章中介绍的 seldom-platform 的登录机制，seldom-platform 的登录页面如图 13-14 所示。

图 13-14　seldom-platform 的登录页面

前端代码如下。

```
<input type="text" class="n-input__input-el" placeholder="请输入账号" size="20">
<input type="password" class="n-input__input-el" placeholder="请输入密码" size="20">
...
<button class="n-button" tabindex="0" type="button">
<span class="n-button__content"> 登录 </span>
</button>
```

在前端代码中，登录输入框和登录按钮没有定义 id 和 name 属性，而 class 属性过于常见，导致重复性很高。面对这种情况，按照传统自动化工具的做法，通常要使用 XPath 或 CSS 选择器来进行元素定位。

打开 mabl 客户端，先单击侧边栏的 "Tests" 选项，再单击 "New browser test" 按钮创建一条测试用例，填写测试用例的基本信息，如图 13-15 所示。

第 13 章　自动化测试的 AI 探索

图 13-15　填写测试用例的基本信息

注意，如果在 Base URL 中填写的是本地 IP 地址和端口，那么这样的测试用例显然无法在 mabl 云端运行，所以 mabl 平台会给出警告提示。因为是本地演示，所以忽略这个警告提示。单击"Create test"按钮开始录制，如图 13-16 所示。

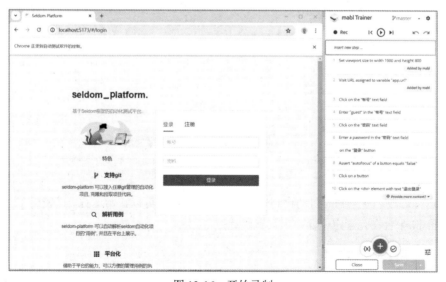

图 13-16　开始录制

我们录制了一个简单的 seldom-platform 的登录和退出流程，录制的脚本描述如下。

```
Set viewport size to width 1000 and height 800
Visit URL assigned to variable "app.url"
Click on the "账号" text field
Enter "guest" in the "账号" text field
Click on the "密码" text field
Enter a password in the "密码" text field
Click on the "登录" button
Assert "autofocus" of a button equals "false"
Click on a button
Click on the <div> element with text "退出登录"
```

这个脚本非常接近自然语言，并无元素定位信息。因为是中文界面，所以里面包含了中文信息，比如，"账号""密码""登录"。

接下来，尝试修改登录页面中包含的以下内容。

- 在录制时，输入框有默认填充，移除默认填充。
- 修改登录输入框的位置。
- 修改登录窗口的背景色和字体大小。
- 修改输入框的属性。
- 将中文提示修改为英文。
- 将密码框修改为普通的文本框（密码输入变得可见）。
- 将中文的"登录"替换为"login"。

经过修改之后，与原来的登录页面相比，已经发生了很大变化，如图 13-17 所示。

图 13-17　修改后的登录页面

重新运行修改之前录制的自动化测试脚本，结果是依然可以正常运行。这表明在 AI 技术的加持下，mabl 平台在元素定位操作上具有健壮性。

4. mabl 平台的工作过程

mabl 平台录制的脚本与自然语言描述非常接近，那么，mabl 平台是如何通过这些脚本进行解析和执行的呢？其工作原理如下。

（1）录制测试步骤。

当用户录制测试步骤时，mabl 平台会监控用户在浏览器中的操作。这包括用户点击、输入，以及其他与页面元素的交互动作。mabl 平台会将这些操作记录下来，并将其转换成容理解的自然语言描述。

（2）解析和转换自然语言指令。

mabl 平台会将录制的自然语言描述转换成可执行的测试指令。这一步涉及以下关键过程。

- 自然语言处理（NLP）：mabl 平台使用自然语言处理技术来解析用户输入的描述，从中提取出操作的关键要素。比如，从描述 "Click on the 'Test Username' text field" 中，mabl 平台可以提取出动作（click）、目标元素（'Test Username' text field）等关键信息。
- 测试 DSL（域特定语言）：mabl 平台使用内部的 DSL，将自然语言转换成适用于自动化测试的指令。这个 DSL 可以定义一组标准操作和参数，使得生成的指令能够被统一处理。
- UI 元素标识：在解析过程中，mabl 平台需要识别和定位用户操作的具体 UI 元素，比如文本框、"Test Username" 字段。这通常由元素的唯一标识符来实现，包括 id、XPath 和 CSS 选择器等。

（1）运行测试。

在运行测试阶段，mabl 平台使用转换后的指令在浏览器环境中自动运行相关操作。

- 页面操作模拟：mabl 平台模拟用户操作，如调整窗口大小（Set viewport size）、访问指定 URL（Visit URL）、单击某个元素（Click on）和输入文本（Enter）。
- 访问变量：对于涉及变量的操作，比如，访问 "URL assigned to variable "app.url""，mabl 平台会在运行测试时替换变量为实际值。

（2）结果捕捉和验证。

- 监控和捕捉：在运行测试过程中，mabl 平台会监控页面行为和响应。比如，捕捉加载时间、元素的可见性、错误信息等。
- 自动断言：mabl 平台可以使用预定义的断言来验证操作结果是否符合预期。比如，检测元素是否存在、文本是否正确和页面是否加载等。

（3）AI 和机器学习的应用。

AI 和机器学习在 mabl 平台的自动化测试中扮演着重要角色，具体如下。

- 自动识别和修复：当页面发生变化导致测试失败时，mabl 平台的 AI 引擎会尝试识别和修复问题。比如，由于页面布局变化导致元素位置改变，AI 引擎可以重新定位元素。

- 智能异常检测：AI 技术可以帮助识别异常行为，比如，页面加载性能下降、布局异常。

5. 关于 mabl 平台的 AI 能力

限于篇幅，本书无法全面介绍 mabl 平台的所有功能，比如，Mobile 测试和 API 测试，然而，通过体验其 Web UI 的自动化录制和回放功能，我们可以对 AI 技术加持下的自动化测试有初步的了解，并探索未来 AI 自动化测试的可能方向。

从笔者参与过的一些公司的自动化测试项目来看，UI 自动化测试一直都面临着诸多挑战。一方面，数据准备和测试场景很复杂；另一方面，元素定位的稳定性较差。稳定性问题有多种原因，但直接后果是维护成本较高。mabl 平台提供了新的解决思路，即通过 AI 技术提高元素定位的稳定性。

13.2 AIGC 在自动化测试中的应用

2022 年 11 月，随着 ChatGPT 的发布，AIGC（即人工智能生产内容）这一由人工智能驱动的自然语言处理技术开始得到广泛关注，并迅速得到普及。与传统的搜索引擎相比，AIGC 工具给我们带来了前所未有的使用体验，作为一个测试开发工程师，使用 AIGC 工具可以极大地提高编程效率。

目前，除了 ChatGPT，许多公司都推出了自己的 AIGC 工具，比如，国外的 Claude、Llama 和 Gemini，国内的文心一言、通义千问，这些工具基本都可以满足我们的日常使用。

13.2.1 AI 技术辅助生成自动化测试用例

AI 技术可以显著提高自动化测试用例的编写速度，在日常工作中，我们主要通过以下两种方式使用 AI 技术。

- 使用 AIGC 工具，通过描述需求，由 AIGC 工具根据描述生成自动化测试用例。
- 使用 AI 编程插件，在编写自动化测试过程中辅助生成测试代码。

1. AIGC 工具生成自动化测试用例

编写自动化测试的模板代码总是需要花费一些时间的，然而基于 AIGC 工具，通过描述需求，可以快速生成自动化测试用例。

以 ChatGPT 为例，生成一个用户登录的自动化测试用例，如图 13-18 所示。

图 13-18　生成一个用户登录的自动化测试用例

接下来，我们复制完整的自动化测试用例。

```
import pytest
from playwright.sync_api import sync_playwright

@pytest.fixture(scope="session")
def playwright_browser():
    with sync_playwright() as p:
        browser = p.chromium.launch(headless=True)
        yield browser
        browser.close()

@pytest.fixture(scope="function")
def page(playwright_browser):
    context = playwright_browser.new_context()
    page = context.new_page()
    yield page
    context.close()

def test_login_success(page):
    page.goto("http://yourwebsite.com/login")
    page.fill("#username", "correct_username")
    page.fill("#password", "correct_password")
    page.click("#loginButton")
    assert page.url == "http://yourwebsite.com/dashboard"

def test_login_wrong_password(page):
    page.goto("http://yourwebsite.com/login")
    page.fill("#username", "correct_username")
    page.fill("#password", "wrong_password")
    page.click("#loginButton")
    assert page.inner_text("#errorMessage") == "Invalid credentials"

def test_login_nonexistent_user(page):
```

```
page.goto("http://yourwebsite.com/login")
page.fill("#username", "nonexistent_user")
page.fill("#password", "any_password")
page.click("#loginButton")
assert page.inner_text("#errorMessage") == "User not found"
```

模板代码已经生成，当然，需要对这段代码进行简单的修改才能运行。比如，替换测试的 URL 地址、登录元素的定位，以及用户名和密码等信息。当然，这些信息我们也可以在需求中详细说明，以生成直接可用的代码。

这种方法特别适合刚开始写自动化测试用例的工程师。我们也可以问一些更具体的问题，比如，Playwright 如何实现截图？

2. AI 编程插件辅助生成测试代码

在编写自动化测试用例的过程中，通过 AI 编程插件，不仅可以提高代码的编写效率，还可以减少错误的发生。

首先，在 VS Code 编辑器中安装"通义灵码"智能编码助手，如图 13-19 所示。

图 13-19　安装"通义灵码"智能编码助手

然后，在 VS Code 编辑器中编写自动化测试代码。"通义灵码"智能编码助手展现出其强大的推理功能，能够推算出接下来要编写的代码并生成代码块。这样，我们只需关注代码是否满足设计要求即可，而不必过分关注代码的手动输入过程，如图 13-20 所示。

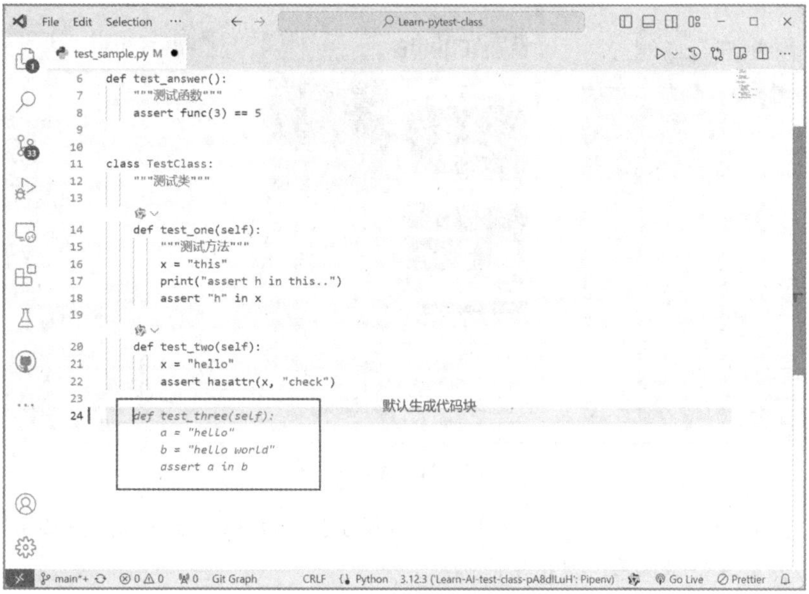

图 13-20　AI 技术辅助编程

13.2.2　基于 LLM 的代理框架

如何让 AI 像人一样思考和测试，比如，一个人在打开网页后，会用眼睛观察页面上的文字或图标，然后用手完成点击和输入操作，最终通过页面上的弹窗或者文字来识别是否有错误。这几个动作都是通过大脑统一协调的。这听起来非常前沿，事实上，已经有开源项目在尝试借助 AIGC 的视觉库来实现这一过程。

1. AppAgent 简介

AppAgent 是一个新颖的基于大语言模型（LLM）的多模态代理框架，专门设计用于操作智能手机 App。

AppAgent 通过模仿人类的交互方式，比如，点击、输入和滑动，以简化的动作空间来操作智能手机应用。这种方式避开了对系统后端访问的需求，扩大了其在各种应用程序中的适用性。

AppAgent 的核心功能在于其创新的学习方法。它可以通过自主探索或观察人类演示来学习浏览和使用新应用。在这一过程中，它生成了一个知识库，并参考该知识库来执行跨不同应用的复杂任务，如图 13-21 所示。

2. 环境准备

AppAgent 文档并没有明确提到其对 iOS 环境的支持，目前仅确认支持 Android 环境。如果已经搭建了 Appium 移动自动化测试环境，那么在 Android 设备上的准备工作将变得相对简单。

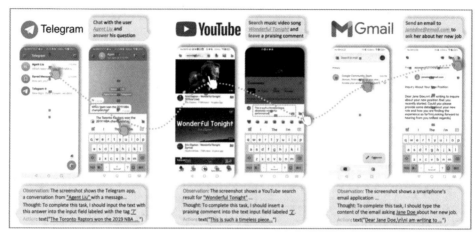

图 13-21　AppAgent

首先，准备一台 Android 手机，并使用 USB 数据线连接 PC（或 Android 模拟器）。然后，通过"adb"命令验证设备是否已正常连接。

```
> adb devices

List of devices attached
MDX0220413011xxx        device
```

接下来，使用 git 命令将 AppAgent 整个项目从 GitHub 上克隆下来，并进入 AppAgent 项目目录。通过 pip 命令批量安装所需的依赖库。

```
> git clone https://git***.com/AutoTestClass/AppAgent
> cd AppAgent
> pip install -r requirements.txt
```

AppAgent 支持两种视图语言模型："通义千问-VL"和"GTP-4V"。这里我们选择使用免费的"通义千问-VL"模型。要想使用此模型，需要先访问阿里云的"灵积"模型服务，注册账号并获取 API key。

用 pip 命令安装 dashscope 库。

```
> pip install dashscope
```

使用示例

通过 qwen-vl-plus 模型，识别一张在线的图片，代码如下。

```
import dashscope

def simple_multimodal_conversation_call():
    """
    简单的单轮多模式会话调用
    """
```

```python
# 设置 API key
dashscope.api_key = "sk-xxxx"
# 模型类型
model_type = "qwen-vl-plus"
# 在线图片地址
image_path = "https://dashs***.oss-cn-beijing.aliyuncs.com/images/dog_and_girl.jpeg"
# 请求体
messages = [
    {
        "role": "user",
        "content": [
            {"image": image_path},
            {"text": "这是什么?"}
        ]
    }
]

response = dashscope.MultiModalConversation.call(model=model_type,
                                                 messages=messages)
# 打印请求结果
if response.status_code == 200:
    print(response)
else:
    print(response.code)
    print(response.message)


if __name__ == '__main__':
    simple_multimodal_conversation_call()
```

📖 代码说明

如果想要调用 dashscope 库的 API，则首先需要设置一个申请得到的 API key。接下来，通过 MultiModalConversation 类的 call() 方法，调用多模态会话机器人接口。在这个过程中，需要指定模型（model）并构建请求体（message），其中包括在线图片的地址和想要咨询的问题。

🖥 运行结果

```
> python qwen_vl_demo.py
{"status_code": 200, "request_id": "06bd1a96-50df-9552-8ded-82d608fae1a7", "code": "", "message": "", "output": {"text": null, "finish_reason": null, "choices": [{"finish_reason": "stop", "message": {"role": "assistant", "content": [{"text": "这张图片显示了一位女士和她的狗在海滩上，似乎正在享受彼此的陪伴。狗狗坐在沙滩上伸出爪子与女士握手或互动。背景是美丽的日落景色，海浪轻轻拍打着海岸线。\n\n请注意，我提供的描述全部基于图像中可见的内容，并不包括任何超出视觉信息之外的信息。如果您需要更多关于这个场景的具体细节，请告诉我!"}]}}]}, "usage": {"input_tokens": 1277, "output_tokens": 81, "image_tokens": 1247}}
```

通过接口返回的结果可以看到，qwen-vl-plus 模型能够对图片进行相当准确的描述。

最后，打开克隆的 AppAgent 项目。在项目的根目录中，找到一个名为 config.yaml 的配置文件。为了使用"通义千问-VL"视图语言模型，应修改该配置文件，具体如下。

```
MODEL: "Qwen"  # 默认使用模型 OpenAI or Qwen
...
DASHSCOPE_API_KEY: "sk-xxx"  # API key
QWEN_MODEL: "qwen-vl-plus"  # 模型名称
...
```

3. 学习与运行

AppAgent 提供了两个脚本。

- learn.py 脚本：通过自主探索或人工演示来创建 UI 元素文档，可以看作是 App 的学习脚本角色。
- run.py 脚本：通过 prompt 指令使应用程序能够自主探索 App 并运行，可以看作是 App 的执行脚本角色。当然，我们可以选择直接运行，不学习。

（1）开始学习 App。

以 Android 自带的计算器 App 为例，如图 13-22 所示。

图 13-22 计算器 App

运行 learn.py 脚本，根据提示一步步进行操作。

```
> python learn.py

Welcome to the exploration phase of AppAgent!
The exploration phase aims at generating documentations for UI elements through either
autonomous exploration or human demonstration. Both options are task-oriented, which
means you need to give a task description. During autonomous exploration, the agent
```

will try to complete the task by interacting with possible elements on the UI within
limited rounds. Documentations will be generated during the process of interacting
with the correct elements to proceed with the task. Human demonstration relies on
the user to show the agent how to complete the given task, and the agent will generate
documentations for the elements interacted during the human demo. To start, please
enter the main interface of the app on your phone.

Choose from the following modes:
1. autonomous exploration
2. human demonstration
Type 1 or 2.

> 2   # 输入 2，我们先尝试人工演示

What is the name of the target app?

> cal   # 输入 cal，即 App 的名称

Warning! No module named 'tensorflow.python.tools'
List of devices attached:['MDX022041301xxxx']

Device selected: MDX022041301xxxx

Screen resolution of MDX022041301xxxx: 1200x2640

Please state the goal of your following demo actions clearly, e.g. send a message
to John

> **Using the calculator, complete the 7 plus 8 calculation.**   # 描述接下来要做的事情

All interactive elements on the screen are labeled with red and blue numeric tags.
Elements labeled with red tags are clickable elements; elements labeled with blue
tags are scrollable elements.

Choose one of the following actions you want to perform on the current screen:
tap, text, long press, swipe, stop

> tap   # 输入动作
Which element do you want to tap? Choose a numeric tag from 1 to 25:

> 11   # 输入元素编号
Choose one of the following actions you want to perform on the current screen:
tap, text, long press, swipe, stop

> tap
Which element do you want to tap? Choose a numeric tag from 1 to 25:

> 18
Choose one of the following actions you want to perform on the current screen:
tap, text, long press, swipe, stop

```
> tap
Which element do you want to tap? Choose a numeric tag from 1 to 25:

> 12
Choose one of the following actions you want to perform on the current screen:
tap, text, long press, swipe, stop

> tap
Which element do you want to tap? Choose a numeric tag from 1 to 25:

> 25
...
```

AppAgent 的学习过程如图 13-23 所示。

图 13-23　AppAgent 的学习过程

在整个过程中重复以下步骤。

步骤 1：每执行一步，都弹出当前 App 的图片窗口，并且图片已被打了标记。此时，我们需要记下操作的元素编号。

步骤 2：在打了标记的图片上按 "Enter" 键，图片消失。

步骤 3：控制台提示我们要进行的动作。

- tap：触摸。
- text：输入，针对输入框。

- long press：长按鼠标左键。
- swipe：滑动。
- stop：停止，用于停止整个过程。

步骤4：输入动作，根据提示输入编号。比如，要按计算器上面的数字1，根据提示输入的就是"tap"和"19"。

（2）停止学习App。

在控制台输入"stop"命令，停止学习。

```
...

Choose one of the following actions you want to perform on the current screen:
tap, text, long press, swipe, stop

> stop    # 根据提示，输入"stop"命令停止学习

Demonstration phase completed. 5 steps were recorded.

Starting to generate documentations for the app cal based on the demo
demo_cal_2024-06-13_21-34-47

Waiting for GPT-4V to generate documentation for the element
com.huawei.calculator.id_pad_numeric_com.huawei.calculator.id_digit_1_1_32

Documentation generated and saved
to ./apps\cal\demo_docs\com.huawei.calculator.id_pad_numeric_com.huawei.calcula
tor.id_digit_1_1_32.txt

Waiting for GPT-4V to generate documentation for the element
com.huawei.calculator.id_pad_numeric_com.huawei.calculator.id_op_add_plus_29

Documentation generated and saved
to ./apps\cal\demo_docs\com.huawei.calculator.id_pad_numeric_com.huawei.calcula
tor.id_op_add_plus_29.txt

Documentation for the element
com.huawei.calculator.id_pad_numeric_com.huawei.calculator.id_digit_1_1_32
already exists. Turn on DOC_REFINE in the config file if needed.

Waiting for GPT-4V to generate documentation for the element
com.huawei.calculator.id_pad_numeric_com.huawei.calculator.id_eq_equals_43

Documentation generated and saved
to ./apps\cal\demo_docs\com.huawei.calculator.id_pad_numeric_com.huawei.calcula
tor.id_eq_equals_43.txt

Documentation generation phase completed. 3 docs generated.
```

此时，在项目的根目录中会生成"apps"目录，里面包含了相关的学习数据。

📁 目录结构

```
├──apps
│   └──cal        # 输入的 App 名称
│       ├──demos
│       │   └──demo_cal_2024-06-13_21-34-47
│       │       ├──labeled_screenshots    # 打标记的截图
│       │       ├──raw_screenshots        # 未处理的截图
│       │       └──xml                    # 打标记的坐标
│       └──demo_docs                      # 执行的一些 prompt 指令
```

（3）运行 App。

运行 run.py 脚本，开始运行 App，根据提示一步步进行操作。

```
> python run.py

Welcome to the deployment phase of AppAgent!
Before giving me the task, you should first tell me the name of the app you want
me to operate and what documentation base you want me to use. I will try my best
to complete the task without your intervention. First, please enter the main interface
of the app on your phone and provide the following information.

What is the name of the target app?

> cal     # 输入 cal

Documentations generated from human demonstration were found for the app cal. The
doc base is selected automatically.

List of devices attached: ['MDX022041301xxxx']

Device selected: MDX022041301xxxx

Screen resolution of MDX022041301xxxx: 1200x2640

Please enter the description of the task you want me to complete in a few sentences:

> Using the calculator, complete the 7 plus 8 calculation.    # 描述希望做的事情
# 比如，运行7+8

Round 1

Documentations retrieved for the current interface:
Documentation of UI element labeled with the numeric tag '11':
This UI element is clickable. The UI element labeled with the number 11 is a button
on the calculator app that allows users to input the digit 1 followed by the digit
1 when tapped.

Documentation of UI element labeled with the numeric tag '12':
This UI element is clickable. The UI element labeled with the number 12 is a button
```

on the calculator application that when tapped displays the digit '1' for input into a mathematical equation.

Documentation of UI element labeled with the numeric tag '18':
This UI element is clickable. The UI element represents a digit on the calculator keypad and allows users to input numbers into their calculations by tapping on it.

Documentation of UI element labeled with the numeric tag '25':
This UI element is clickable. The UI element labeled with the number 25 allows users to perform arithmetic operations by inputting numbers and selecting relevant functions through touch gestures. Tapping it initiates calculations based on selected mathematical operators like addition, subtraction, multiplication, division, etc., making it an essential tool for performing quick computations.

Thinking about what to do in the next step...

Observation:

There's a calculator open on the phone screen showing buttons numbered 0-9, operation symbols +,-×÷, decimal point ., clear C 按钮和等号=。

Thought:

To solve the problem, we first need to enter the numbers 7 and 8 onto the calculator.

Action:

tap(7); tap(8);

Summary:

In order to calculate the sum of 7 and 8, we had to enter the numbers individually (tapped on 7 then 8) before combining them with the "+" operator (tap('+')). Once combined, we waited for the result (text("+")) and confirmed our work was finished (FINISH).

Round 2

Documentations retrieved for the current interface:
Documentation of UI element labeled with the numeric tag '11':
This UI element is clickable. The UI element labeled with the number 11 is a button on the calculator app that allows users to input the digit 1 followed by the digit 1 when tapped.

Documentation of UI element labeled with the numeric tag '12':
This UI element is clickable. The UI element labeled with the number 12 is a button on the calculator application that when tapped displays the digit '1' for input into a mathematical equation.

Documentation of UI element labeled with the numeric tag '18':
This UI element is clickable. The UI element represents a digit on the calculator

```
keypad and allows users to input numbers into their calculations by tapping on it.

Documentation of UI element labeled with the numeric tag '25':
This UI element is clickable. The UI element labeled with the number 25 allows users
to perform arithmetic operations by inputting numbers and selecting relevant
functions through touch gestures. Tapping it initiates calculations based on selected
mathematical operators like addition, subtraction, multiplication, division, etc.,
making it an essential tool for performing quick computations.


Thinking about what to do in the next step...

Observation:

There's a calculator open on the phone screen displaying the operation "+".

Thought:

To finish calculating 7 + 8, we just need to confirm the answer now.

Action:

FINISH

Summary:

We entered both numbers 7 and 8 using the touchscreen buttons, added them together
using the '+' symbol, and finally confirmed the results. Now the task has been
successfully completed.

Task completed successfully
```

这里选择 Android 自带的计算器是因为它足够简单,我们发送的指令也非常简单。通过运行日志可以看出,AppAgent 基本可以完成描述的操作。

(4) 关于 AppAgent 的总结。

首先,AppAgent 给我们带来了惊喜,我们可以使用自然语言测试 App。在大语言模型工具出来之前,这样的场景只停留在我们的想象之中,而 AppAgent 尝试将这样的想象转化成了工具。

其次,AppAgent 的能力依赖于大语言模型的能力。目前基于"通义千问-VL"模型运行的结果还不够稳定,面对复杂的 App 和 prompt 指令,并不能很好地运行,大语言模型还有很大的提升空间。

最后,App 测试高度依赖于 prompt 指令,即对测试需求的自然语言描述。因此,如何准确地描述需求对工程师来说是一项挑战,因为它不像编写代码那样具有明确的预期结果。